Ecologies and Economies in Medieval and Early Modern Europe

# Brill's Series in the History of the Environment

*General Editor*

Aleks Pluskowski, University of Reading

VOLUME 1

Richard C. Hoffmann

# Ecologies and Economies in Medieval and Early Modern Europe

## Studies in Environmental History for Richard C. Hoffmann

*Edited by*

Scott G. Bruce

BRILL

LEIDEN • BOSTON
2010

*On the cover*: Woodcut from Jacob Froelich's *Ein Wunder Kuenstreiches Buechlein...*, 1531. Otto von Kienbusch Angling Collection. Rare Books Division. Department of Rare Books and Special Collections. Princeton University Library. With kind permission of Princeton University Library.

This book is printed on acid-free paper.

Library of Congress Cataloging-in-Publication Data

Ecologies and economies in medieval and early modern Europe : studies in environmental history for Richard C. Hoffmann / edited by Scott G. Bruce.
p. cm. -- (Brill's series in the history of the environment)
Based on a conference entitled "Lakes, Rivers and Fish: A Celebration of Richard Hoffmann's Work," held on May 4, 2009, at York University in Toronto.
Includes index.
ISBN 978-90-04-18007-9 (hardback : alk. paper) 1. Human ecology--Europe--History. 2. Nature--Effect of human beings on--Europe--History. 3. Biotic communities--Europe--History. 4. Fishes--Ecology--Europe--History. 5. Lake ecology--Europe--History. 6. Civilization, Medieval. 7. Hoffmann, Richard C. I. Bruce, Scott G. (Scott Gordon), 1967- II. Title. III. Series.

GF540.E27 2010
304.2094'0902--dc22

2009039920

ISSN 1876-6595
ISBN 978 90 04 18007 9



PRINTED IN THE NETHERLANDS

CONTENTS

## ACKNOWLEDGMENTS

A volume of this kind succeeds due to the generosity of its contributors. I am very grateful to everyone who so graciously shared their work to honor the influence of Rich's career. On 4 May 2009, many of us gathered at York University in Toronto for a conference entitled "Lakes, Rivers and Fish: A Celebration of Richard Hoffmann's Work," where historians of medieval Europe and modern Canada convened for a day of papers and discussion about common themes in environmental history. Many thanks to Myra Rutherdale and Jean Levy for organizing the conference; to Nick Rogers, Jonathan Edmondson, and the Department of History for their support; to York University's Avie Bennett Chair in Canadian History, The Nelles Environmental History Endowment, and NICHE (Network in Canadian History and Environment) for their generous sponsorship of the event; to Connie Berman, Maryanne Kowaleski and Paolo Squatriti for attending and presenting their contributions for this volume; to the Canadian historians who shared their research in the afternoon session: H. V. Nelles, Matthew Evenden, and Elizabeth Piper; and to Richard Unger for agreeing to give the 2009 H. V. Nelles Lecture in Environmental History for this occasion. I am also particularly grateful to Ellen Hoffmann, who has been a silent and very helpful partner in this project since its inception.

The final work on the manuscript of this volume took place in June 2009 while I was a Visiting Fellow of *Conventus*, a European research collective on problems related to religious communal life in the Middle Ages, based at the Department of Medieval History at the University of Ghent in Belgium. I am indebted to Professor Steven Vanderputten for the invitation to come to Ghent and for the precious research time afforded by the generous fellowship.

SGB
Ghent, Belgium

## LIST OF FIGURES AND TABLES

*Kowaleski*



*Reynard*

*Van Neer and Ervynck*

## LIST OF CONTRIBUTORS

*Constance H. Berman* is Professor of History at the University of Iowa, Iowa City, where she has taught for twenty years. She is the author and editor of numerous books, including *Medieval Religion: New Approaches* (Routledge, 2005); *The Cistercian Evolution: The Invention of a Religious Order in Twelfth-Century Europe* (University of Pennsylvania Press, 2000); and *Medieval Agriculture, the Southern French Countryside, and the Early Cistercians* (American Philosophical Society, 1986). She is working on land, water, mills, and villages in southern France as part of a new project, *Women's Work in the European Economic Expansion, 1000–1350 AD.*

*Scott G. Bruce* is Associate Professor of History at the University of Colorado at Boulder. He traces his academic roots to the Department of History at York University, where he studied as an undergraduate under Rich Hoffmann. His research focuses on the abbey of Cluny from the tenth through the twelfth centuries. He is the author of *Silence and Sign Language in Medieval Monasticism: The Cluniac Tradition, c. 900–1200* (Cambridge University Press, 2007), a topic suggested to him by Rich. He is currently writing a study of western perceptions of Islam in Cluniac hagiography.

*Anton Ervynck* coordinates a team of natural scientists at the Flemish Heritage Institute (Belgium). His main research area is the analysis of animal remains from premodern Flemish sites. His primary interests are the study of social differentiation through consumption patterns, and the history of natural and cultural environments. His recent published work (many co-authored and co-edited) includes: "Beyond Affluence: The Zooarchaeology of Luxury," *World Archaeology* 34 (2003), 428–41; "*Orant, pugnant, laborant*: The Diet of the Three Orders in the Feudal Society of Medieval North-Western Europe," in *Behaviour Behind Bones: The Zooarchaeology of Ritual, Religion, Status and Identity*, ed. S. J. O'Day, W. Van Neer and A. Ervynck (Oxbow Books, 2004); and *Pigs and Humans: 10,000 Years of Interaction* (Oxford University Press, 2007).

*William Chester Jordan* is Dayton-Stockton Professor of History and Chairman of the History Department of Princeton University. He is the author of many books and articles, most recently *A Tale of Two Monasteries: Westminster and Saint-Denis in the Thirteenth Century* (Princeton University Press, 2009). His current research focuses on English-French relations in the thirteenth and early fourteenth centuries.

*Maryanne Kowaleski* is Joseph Fitzpatrick S.J. Distinguished Professor and Director of Medieval Studies at Fordham University. She is the author and editor of numerous books and articles on women and work, the urban family, port towns, and overseas and inland trade in medieval England, including *Local Markets and Regional Trade in Medieval Exeter* (Cambridge University Press, 1995); *Gendering the Master Narrative* (Cornell University Press, 2003); and *Medieval Towns: A Reader* (Broadview Press, 2006).

*Pierre Claude Reynard* is Associate Professor in the Department of History at the University of Western Ontario, Canada, with an interest in the social, economic, and environmental history of early modern and modern France and Europe. His recent work has focussed on preindustrial manufacturing, its techniques, strategies, and relations with the state, as well as urban expansion and entrepreneurship. He also has a broad interest in environmental history ranging from regulations and public works to the place of rivers in the lives of communities.

*Paolo Squatriti* teaches medieval European history at the University of Michigan, where he specializes in early medieval Italian history. Most recently he was the editor of the collection *Natures Past: The Environment in Human History* (University of Michigan Press, 2007) and the translator of *The Complete Works of Liudprand of Cremona* (Catholic University of America Press, 2007). He likes both the tree and the nut of *Castanea sativa* and is now working on the early medieval history of the chestnut in Europe.

*Richard W. Unger*, a fellow student in the Yale doctoral program with Richard Hoffmann, is a professor in the Department of History of the University of British Columbia. He teaches medieval and economic history and has published on the history of technology, specifically on the evolution of European ship design from the classical world through to the era of steam and on the development of the Dutch and European brewing industries. His most recent books are *Beer in the Middle Ages and the Renaissance* (University of Pennsylvania Press, 2004) and *Pictures of Power: Ships on Maps in Renaissance Europe* (forthcoming).

*Petra J. E. M. van Dam* holds the chair for water history at the Vrije Universiteit, Amsterdam. She specializes in the pre-industrial environmental history of northwestern Europe, in particular on landscape transformations in aquatic and terrestrial eco-systems. A recent example of this research (co-authored with Milja van Tielhof) is: "Losing Land, Gaining Water: Ecological and Financial Aspects of Regional Water Management in Rijnland, 1200–1800," *Water Management, Communities and Environment: The Low Countries in Comparative Perspective, c.1000–c.1800, Jaarboek voor Ecologische Geschiedenis 2005/2006* 10 (2006), 63–94. Professor van Dam is currently finishing a monograph on the history of humans and animals in the coastal sand dunes of Holland.

*Wim Van Neer* is the head of the Bioarchaeology Unit at the Royal Belgian Institute of Natural Sciences in Brussels, Belgium, and teaches archaeozoology at the Katholieke Universiteit Leuven. He conducts faunal analysis in Europe, the Near East and Africa and has a strong interest in the study of fish remains from archaeological sites. His recent publications (some collaborative and co-authored) include: "Evolution of Prehistoric Fishing in the Nile Valley," *Journal of African Archaeology* 2 (2004), 251–69; "The Emergence of Fishing Communities in the Eastern Mediterranean Region: A Survey of Evidence from Pre- and Protohistoric Periods," *Paléorient* 31 (2005), 131–57; and "Salted Fish Products from the Coptic Monastery at Bawit, Egypt: Evidence from the Bones and Texts," in *The Role of Fish in Ancient Times*, ed. H. Hüster Plogmann (Verlag Marie Leidorf, 2007), pp. 147–59.

*Verena Winiwarter* is professor of Environmental History at Alpen-Adria University in Klagenfurt, Austria. Her degrees are in technical chemistry, history, and human ecology. Together with Martin Knoll, she has published the first German textbook for environmental history: *Umweltgeschichte: Eine Einführung* (Böhlau UTB, 2007). Her research interests include the history of soils and conceptual issues in doing and teaching environmental history.

# PREFACE: RICHARD C. HOFFMANN, AN APPRECIATION

There is something down there and
you want it told.
Gwendolyn MacEwen (1941–1987),
*Dark Pines Under Water*

Those who know Professor Richard C. Hoffmann know all about his relentless pursuit of aquatic life in the rivers and coastal waters of premodern Europe. Few of us can boast that we love our work as much as he does. An avid angler in American and European rivers, Rich has applied all of the joy and enthusiasm of this pursuit to the formidable body of award-winning scholarship that he has produced on early fish husbandry, the history of fishing, and the impact of human beings on the premodern environment. On these topics, he has been a pathbreaking explorer. Beginning in 1985 with the appearance in *Speculum* of an article entitled "Fishing for Sport in Medieval Europe: New Evidence" and followed closely by an unprecedented treatment of medieval fishponds, written at the invitation of Joseph Strayer for the *Dictionary of the Middle Ages*, Rich began changing the way that historians think about the interaction between medieval people and the natural world in which they lived.[1] Many studies followed from these early works, most notably "Economic Development and Aquatic Ecosystems in Medieval Europe," which appeared in *The American Historical Review* in 1996. This article brought the study of medieval environmental history to the broadest possible audience of professional historians and won the prestigious Alice Hamilton Prize (1997) for best article in the field of environmental history from the American Society for Environmental History. In all of his published work, Rich has harnessed the energy of a life-long passion for fishing to a profound sympathy for the mutual influence of natural phenomena and human activity in the premodern world. The latter quality was already evident

---

[1] For a complete bibliography of Rich Hoffmann's published works to date (2008), see Appendix A.

in one of his earliest published works, an article entitled "Warfare, Weather and a Rural Economy: The Duchy of Wrocław in the Mid-Fifteenth Century," which added evidence of hostile environmental conditions to the list of causes behind the agricultural crisis that struck Silesia in the later Middle Ages. But fishing was never far from Rich's mind, even while he was writing his dissertation on medieval Poland in New Haven. He once told me that as a graduate student he employed a swivel chair at his desk with a dual purpose in mind. When he needed a break from pounding out the dissertation on his typewriter, a simple 90-degree turn directed his attention to the thread, hackles and tools of the fly fisherman that he always kept by his side. When he tired of tying flies, another turn of the chair returned his attention to the distant past. Like many American midwesterners, who grow up in a world of rivers and lakes, Rich was and remains "haunted by water."[2]

Not every famous explorer is also a successful cartographer, who can chart a clear and inviting path through otherwise difficult territory and thereby encourage others to follow his lead, but in his long career Rich Hoffmann has been both to many colleagues and students. Through his tireless sponsorship of sessions on premodern environmental history at the International Medieval Congress in Kalamazoo, Michigan, Rich has provided a unique forum for North American and European historians of all backgrounds to present their work on every conceivable aspect of human interaction with the natural world in medieval and early modern Europe. Participants in these sessions, several of whom are represented in the present volume, always found a welcoming audience of like-minded scholars open to interdisciplinary approaches to Europe's past. One of the biggest draws, many have told me, was firsthand access to Rich's expertise on these subjects, which he shared by means of formal and incisive comments on papers (enough to fill a volume on their own) and informal advice, imparted over a meal or a beer, that tapped the rich vein of his knowledge of the historiography of the premodern natural world in a staggering number of scholarly languages.

As daunting and esoteric as much of this historiography can be, Rich has the rare and enviable talent of making his field of expertise accessible and exciting to his students. In 1992/93, I was a participant in his

---

[2] Norman F. Maclean, *A River Runs Through It and Other Stories*, 25th Anniversary Edition (Chicago and London, 2001), p. 104.

first undergraduate seminar on the topic of premodern environmental history at York University and recall distinctly the sense of standing on an intellectual precipice and looking down onto the canopy of an ancient forest of new possibilities. Rich already knew what was down there, but he shared it with us as though it was new to him as well, giving us all the sense that we were charting this territory together, even though the path had already been prepared by his labor. I had to become a teacher myself before I realized how generous this was of him. As a student I was too caught up by Rich's contagious excitement, in my first reading of Glacken's *Traces on the Rhodian Shore* and Crosby's *Ecological Imperialism*, and in the possibilities that his syllabus offered, in all of its bibliographical richness. Almost twenty years later, a perusal of the original syllabus for this course still evokes the feeling of intellectual discovery that is experienced at the start of each new semester by those who have been fortunate enough to call Rich their teacher.

* * *

The papers collected in this volume represent as much in their formidable erudition as in their dazzling diversity the integrity and vibrancy of the field of premodern environmental history that Rich's scholarship has done so much to foster. The collection is organized into two thematic sections, both of which entertain long chronological views. The first, "Premodern People and the Natural World," comprises papers that explore the complex relationships between human beings and the world in which they lived. While medieval and early modern people were willing and able to modify and control natural resources in creative ways to increase their productivity, these studies underscore the fact that human ingenuity was often a response to the realities of hunger and its allied hardships in preindustrial societies. Far from being carefree exploiters of the earth's riches, premodern people were constantly adapting their means of production and cultivation in the face of a natural world that they could not hope to control for long. The second section of papers, "Aquatic Ecosystems and Human Economies," continues the themes of the first, but focuses specifically on the relationship between human beings and the watersheds that surrounded and sustained them. More than any other historian, Rich has pioneered the research in this specific field of inquiry. It is only fitting that studies of the still pools, fast-flowing rivers and deep-sea currents of European waterways, with their abundance of fauna and indisputable impact on human economies, find their place in a collection of papers in honor of his scholarly achievements.

As the contents of this collection make clear, medieval and early modern environmental history has taken on a coherent shape in recent decades, in no small part due to the efforts of Rich and his collaborators. While the papers in this volume highlight the variety of topics that beckon historians engaged in the study of Europe's natural past, they nonetheless share a broad interdisciplinary methodology that informs their approach to premodern environmental history. The integration of evidence derived from historical, archaeological and natural scientific data is a hallmark of this approach and a recent achievement that Rich's scholarship has done much to foster. This interdisciplinary methodology also provides a framework of understanding between premodern historians and those specialists studying human impact on the natural economy in the post-industrial world. Camped on common ground, environmental historians of all stripes have found that they share many of the same methodological concerns in their attempts to infer the character of past societies and their relationship with the natural world. The ensuing dialogue between specialists across the historical spectrum has enlivened and enriched this field of inquiry. The discipline of environmental history is no stagnant backwater; it is very much a welcoming harbor busy with the commerce of fresh ideas and open to the possibility of new perspectives.

After an introduction by Richard W. Unger that situates the work of Richard Hoffmann in the *longue durée* of the discipline of environmental history, our collection begins at the turn of the first millenium in the woodlands of central Italy. In recent years, many historians have been drawn to the study of premodern forests, both with respect to their role in local economies and their impact on the human imagination. Using the evidence of charters, Paolo Squatriti makes the case for the centrality of an over-looked forest commodity along the Amalfi coast: the chestnut. Far from being isolated holdovers from classical forests on the fringes of cleared and cultivated lands, Squatriti argues that chestnut groves drew the commercial interest of wealthy landowners because of their sustainable productivity (enhanced by their remarkable amenability to human care) and their value as a cash crop in the seaborne commerce of the Mediterranean Sea. He explains how laborers tended, cultivated, and prepared chestnuts for export, thus bringing to light the sophistication of early medieval arboricultural practices and thereby restoring the chestnut tree to its rightful place at the center of a carefully managed Italian woodland that played an important role in a thriving early medieval economy.

For this occasion, William Chester Jordan revisits the topic of his award-winning book, *The Great Famine: Northern Europe in the Early Fourteenth Century*.[3] Between 1315 and 1322, unusually harsh winters and summers of relentless rain cast a pall over the face of northern Europe and led to the most sustained subsistence crisis in medieval history. Jordan considers new sources and studies that have come to light since the publication of his book on the Great Famine in 1996. His paper reevaluates the causes of the famine, the experience of medieval people caught in the misery it created, and the consequences of this catastrope, particularly its relationship to the devastating plague cycle known as the Black Death that struck northern Europe only a generation later (1348–1351). These new findings corroborate the fact that the Great Famine was a natural calamity of unprecedented severity that touched the lives of millions of people across late medieval Europe.

Petra van Dam's paper introduces us to some very enterprising rabbits. In the early modern period, the management of rabbit warrens in the coastal dunes of Holland was a serious preoccupation for local counts, who owned and systematically exploited rabbit populations for their fur and meat. It was forbidden by law to kill rabbits beyond this annual cull, so farmers and gardeners on the dunes had to take defensive measures to protect their produce from unwelcome intrusion by erecting elaborate fences and digging formidable ditches (sometimes in combination). By examining the use of these "micro-environmental intrastructures" in the context of dramatic changes in land use along the dunes between 1400 and 1700, Van Dam shows that farmers and gardeners constructed these complex and expensive barriers with a keen awareness of rabbit behavior based on observation of their life cycle and grazing activities.

Verena Winiwarter's contribution reveals medieval and early modern agronomists at work on the question of soil fertility. What role did the soil play in the growth of plants? Were some kinds of soil more conducive to plant fertility than others? What could human beings do to modify and improve soil quality in order to achieve long-term soil sustainability? These questions are commonly attributed to

[3] William Chester Jordan, *The Great Famine: Northern Europe in the Early Fourteenth Century* (Princeton, N.J., 1996). In 2000, this book won the Haskins Medal from the Medieval Academy of America.

Enlightenment-era scientists, long considered the pioneers of this particular branch of earth science. Winiwarter challenges this historiographical presumption by examining the traditional ecological knowledge imbedded in a wide range of premodern agricultural treatises, from the tenth-century *Geoponika*, a Byzantine collection of agrarian wisdom commissioned by Emperor Constantine VII Porphyrogenitus, to John Evelyn's *Terra: A Philosophical Discourse of the Earth* (1675) and Johann Theodor Eller's *Recherches sur la fertilité de la terre général* (1749). While these works vary considerably with respect to their context and the sensibilities and theoretical approaches of their authors, Winiwarter argues that they all have in common a shared body of practical, experience-based knowledge about soil constituents and the benefits of human intervention in its quality through the application of manures and composts.

The second half of the volume begins with Maryanne Kowaleski's innovative study of the thriving fishing industries of medieval England, where the seasonal migrations of fish to their spawning grounds had a profound influence not only on the orchestration of regional markets but also on the lives of fishermen and their families. Kowaleski charts the migration of herring in the North Sea and the fishers and traders who followed them. The economic impact of their migrations is most apparent in the dates of seasonal fairs in towns along the eastern coast of England, which coincided with the herring harvest. Kowaleski then broadens her discussion to consider the emergent western fisheries off the coasts of Cornwall and Devon and the entry of British fishermen into Atlantic maritime economies as far afield as Iceland, Greenland and Newfoundland. The seasonal absence of fishermen, often for months at a time, had a powerful effect on the rhythms of daily life in fishing towns, where mariners' wives had to manage households without their husbands, and did not go unnoticed by policy makers, who issued special dispensations for the activity of foreign fishermen in their waters and passed legislation to protect important species from over-fishing. Kowaleski argues that the commercial diaspora of English fishermen to distant waters in the later Middle Ages led to advances in curing technologies that made these new fisheries commercially viable. The availability of increased quantities of cured fish did little, however, to alter the rhythm of English fishermen and traders, who continued to organize their industry around the natural migrations of their catch.

Constance Berman's note examines records from the hospital of Saint Thomas at la Trinquetaille near Arles in the Rhône delta. In the decades around 1200, she finds that families owing rents to the hospital often squared their debts to the community with an unusual form of currency: eels. Like other aquatic wildlife, eels were exempt from Christian fast restrictions, which applied primarily to the flesh of quadrupeds. They were thus in high demand in religious communities, especially during Lent. Berman argues that the eels of la Trinquetaille were captured and perhaps even cultivated in enclosed ponds in the marshy Camargue, where generations of human interaction with the Rhône watershed through land clearance, swamp drainage, and the planting of trees and vines inadvertantly created localized habitats that were favorable to the propagation of eel populations, which mature in fresh water before returning to the sea to spawn.

Pierre Claude Reynard's paper examines the debates engendered in the eighteenth century by proposals to expand the city of Lyon beyond what many contemporaries considered to be its natural boundary: the Rhône river. Perched on the bank of the wide and fast-flowing Rhône and hemmed in by steep hills, Lyon is an old city, boasting a Roman and medieval pedigree. By the 1730s, unprecedented population growth finally betrayed the topographical constraints that had always kept the city limits in check. As Reynard shows, the practical difficulties presented by a cramped and crowded urban population colluded with Enlightenment-era visions of more open, and therefore more healthy, cities. Over the course of the eighteenth century, conflict brewed between those whose professional and political interests favored the expansion of Lyon across the Rhône and those who resisted this rationalization of natural space for the sake of urban improvement. Reynard charts how these citizens imagined and evoked the river in their debates surrounding four major renewal projects that transformed the urban landscape of Lyon in the decades around 1750.

Lastly, the collaborative contribution of Wim Van Neer and Anton Ervynck holds a special place in this volume because it illuminates how premodern historical research on the natural world can influence and benefit modern policy-makers in their decisions about crucial environmental issues. Taking as their case study the Water Framework Directive (WFD) established by the European Union in 2000 for the maintenance of water quality, Van Neer and Ervynck argue that historical and archaeozoological evidence from the premodern period provides an important but overlooked indication of the abundance of fish and the biodiversity of river systems before the impact of the Industrial Revolution. Although these data are often fragmentary, they can nonetheless factor usefully into the formulation of "natural" target conditions for regulating authorities like the WFD. Taken together, documentary and archaeological research will add significantly to the historical reconstruction of environmental conditions before the periods of the most significant and sustained human disturbance. In the hands of responsible policy-makers, the results of scholarly inquiry into the quality and character of the premodern environment may one day play a significant role as benchmarks for the restoration, maintenance and protection of modern ecosystems in polities across the globe.

Scott G. Bruce
University of Colorado at Boulder

# INTRODUCTION: HOFFMANN IN THE HISTORIOGRAPHY OF ENVIRONMENTAL HISTORY

Richard W. Unger

Pre-modern environmental history has taken on a more consistent shape in the time since Richard Hoffmann became a practicing historian. The expansion and growth of the field has in no small part been due to his work but, more importantly, he has left his mark on the emerging parameters of a discipline now recognized for its topics, methods, and products. It is only logical that environmental history, like other fields within the larger scope of the study of the past, should pass from a pioneer to a building to an established phase. Richard Hoffmann has been a part of each and in the process has used the amorphous and highly fluid nature of a nascent discipline to leave his extremely positive stamp on what should be done to advance the field. The early development of environmental history, coming in a time marked by the "me" generation, has spawned extensive self-examination.[1] At best this has become self-criticism leading to improvement in the practice of the trade. More often, as here, the tendency has been to rehearse the same or similar themes. At the least, though, such expression of reflection about what environmental history is has allowed a young and vibrant discipline to achieve maturity.

Scholars in the United States took the lead in the formation of environmental history. The need to examine the ways people treated their surroundings was not an idea borrowed from European scholars. Political, economic, social, and other types of history in the New World were often modeled on practices in the Old World. Environmental

[1] See, for example, the essays presented under the theme "Theories of Environmental History," in *Environmental Review* 11 (1987); Alfred W. Crosby, "The Past and Present of Environmental History," *American Historical Review* 100 (1995), 1177–89; J. R. McNeill, "Observations on the Nature and Culture of Environmental History," *History and Theory* 42 (2003), 5–43; and the extremely informative "AHR Conversation: Environmental Historians and Environmental Crisis," *American Historical Review* 113 (2008), 1431–65, to which Richard Hoffmann was a principal contributor.

history was exceptional in that way and so bore the stamp of the struggle in the minds of Americans about how exceptional their society, created over the previous two hundred and more years, actually was. The discussion about the results of Columbus' first voyage of discovery, given new life by the muted celebrations surrounding the 1992 quincentenary of navigation to a land mass then unknown to Europeans, was an obvious case where evaluation depended on the sense of America being unique. That in turn had an effect on evaluating the ensuing changes in the environments of both Europe and the Americas. The tendency for many early scholars who took up environmental questions to see them in terms of an America that suffered damage from the hands of Europeans has, however, been less of a factor in the study of the pre-modern than the modern past. Exceptions exist, of course, such as the history of the British North American colonies before 1776, but in general pre-modern environmental history has been less tied to destruction of some ideal past, a central theme in American environmentalism. This has created a productive tension in the field and in the minds of its practitioners.

Environmental history is both new and old. Pre-modern environmental history has benefitted from the general interest in ecology that has emerged in the last half century. The 1950s saw the first burst of a growing interest in protecting the environment.[2] Fears aroused about damage to wildlife and the land, some of which had already emerged even in the 1920s in public policy debates about soil erosion, were swamped by pressing global political conflicts in the 1940s. Those fears took on a new urgency after World War II. A whole range of questions about the interaction of humans with the world around them can be lumped into a single category. Contemporary environmental issues captured the popular imagination especially in the United States. This concern with ecology has gained momentum ever since and it is certainly not surprising that the interest should generate a search in the distant past to understand the long-term relationship of people with the environment. Driven by present-day problems, the study of pre-modern environmental history also has deep roots in historical work, especially

[2] On the 1950s as the point when the environment came under siege and protection of the environment became critical, see Christian Pfister, "Das '1950er Syndrom': Die umweltgeschichtliche Epochenschwelle zwischen Industriegesellschaft und Konsumgesellschaft," in *Das 1950er Syndrom: Der Weg in die Konsumgesellschaft*, ed. Christian Pfister (Bern, 1995), pp. 51–95.

local history, that had been going on before the conception of the subfield of the discipline. There was much history done before the mid-twentieth century that was environmental even though the practitioners did not know that that was what they were doing. Just as seventeenth-century commentators, worried about the depletion of forests, did not know that they were environmentalists, so too historians who dealt with environmental topics did not know they were part of a new kind of work in the discipline. It took a little time before their work could be placed into a recognizable category along with other studies with dissimilar methods, goals, and approaches but which dealt with similar issues of relationships among living species and the physical world.[3]

Since there were many older studies in varied forms, there were many different roots of pre-modern environmental history. Richard Hoffmann is a classic case of the evolution of the field. He started his work before there were environmental historians. In the spirit of the *Annales* school, he brought to bear many sub-disciplines of history and a wide range of skills to try to appreciate the functioning over a lengthy period of time of a society in one region. His award-winning book on the Duchy of Wrocław over some two centuries marshaled the tools of demography, economics, and politics to lay out and gain some sense of the development of an agrarian regime under pressures that were in part environmental. That is not to say that aspects of country life involving other species and physical surroundings escaped notice. There were in the book at least thirteen references to fish and fishing, many of those a by-product of the study of lordship, and there was also an earlier article on weather and the rural economy of the region.[4] His study of late medieval Silesia indicated a major source for pre-modern environmental history and some of the constraints from which it would emerge.

Economic history was almost by definition concerned with environmental questions. While those concerns were not always a central topic, the influence of short- and long-term changes in the world around them were an off-stage presence for those writing about the economy. That was especially true of pre-modern economic history where the margins of surplus were small or nonexistent. The study of the material

[3] On forest awareness, see Joachim Radkau, *Nature and Power: A Global History of the Environment*, trans. Thomas Dunlap (New York, 2008), pp. 139–42.

[4] Richard C. Hoffmann, *Land, Liberties, and Lordship in a Late Medieval Countryside: Agrarian Structures and Change in the Duchy of Wrocław* (Philadelphia, 1989); and idem, "Warfare, Weather, and a Rural Economy: The Duchy of Wrocław in the Mid-Fifteenth Century," *Viator* 4 (1973), 273–305.

culture of the past, an offshoot to some degree of the *Annales* school, again by definition was also involved in examining environmental matters since the objects involved were the tools, large and small, that people interposed between themselves and the world around them. The same needs applied to the emerging field of the history of technology, devoted as it is to studying the creations of people to deal with their circumstances. The need for government policies to deal with the environment are not unique to recent history. Often political jurisdictions in the distant past had to deal with environmental questions such as competition among different users of a common resource. Political history then, though admittedly often incidentally, had to do with ecological matters. The nostalgia for the past that grew up among social historians also led to a discussion of environmental issues. The musings on a world that has been lost in the era of rapid industrial and population growth over the last two centuries were again almost automatically committed to a concern for the differences between treatment of the environment in the pre-modern era and in the present.[5] Whether the central subject for the social historian might be childhood, country life, or society in general, the examination of a lost and better world implied at the very least in passing a study of people and the world around them. Scientists as well found themselves concerned with pre-modern environmental history. As the field of ecology grew and the study of the contemporary environment expanded, researchers doing the measuring realized they needed yardsticks, base-lines against which to measure their findings. Scientists were then driven to look at time series to assess deviation from some historical norms or to assess any trends they might notice. Most obvious with studies of climate, the drive to find data on which to base judgments about the contemporary environment pushed research back further in time.[6]

While the study of the pre-modern environment was well underway in a number of different and disparate forms by the 1950s, not all of these roots of the discipline of environmental history shared the same background. Different training was needed for each of the different approaches and there were always deficiencies among practitioners, the scientists for example knowing little about historical methods and the humanists among the historians woefully ignorant of the

[5] The obvious example is Peter Laslett, *The World We Have Lost* (London, 1965).
[6] See the contribution of Van Neer and Ervynck in this volume.

science involved. The varied forms of work created the potential for conflict and almost certainly the misunderstanding of findings. What all these investigators shared was an engagement with the environment. Building on the many roots of the field, in the last half century pre-modern environmental historians have learned to exploit these approaches and to create from them a discipline which is broadly based and requires command of a wide range of methods and skills among those consciously working in the sub-field. Richard Hoffmann is not alone as a promoter of using what scientists can teach. He, like others, is aware of the problems of gaining the required knowledge not only because of the need for training in a number of different skills, but also because of the rapid development of ecology as a science, generating new methods and data at an ever-increasing rate, which turns the search for knowledge of the field into a chase of a moving target. He is also aware of a reluctance among historians and the general public to engage with numbers, science and analysis in general, an attribute that continues to challenge pre-modern environmental history.

Richard Hoffmann has joined others as a proponent of training historians in the necessary sub-disciplines or related disciplines. The new generation of scholars should be those who, by the character of their training, can escape from the roots of environmental history in all the other fields of historical study. In the process, environmental history has continued to create its own character with its own body of common knowledge. New ways of doing environmental history are springing from that effort. It is also generating institutions, such as the American Society for Environmental History, the European Society for Environmental History and a World Congress of Environmental History, to promote and support the diverse methods and topics that are now identified as part of the field. The emergence of new ways of going about studying the complex and ever-changing relationships between humans and the world in which they lived during the pre-modern era has generated a thriving and evolving field, which can now make considerable contributions to an understanding of the distant past in all parts of the world. It can also force historians in general to pay attention to environmental factors in the discussion of all aspects of the past.

The shape of pre-modern environmental history has in part been dictated by the pursuit of novel findings, no matter the source. There has been a strenuous effort to ferret out from past historical work any aspects that might contribute to the study of the environment along with a search for sources that, either previously ignored or used for other purposes, might add to the body of knowledge about humans and their surroundings. In this, Richard Hoffmann again offers a classic example of the approach toward novelty and the positive results that flow from it. He has consistently been a proponent of pursuing new findings and incorporating the work of others. This merging of the roots of environmental history with otherwise neglected sources has set the field off in a number of varied directions. It has also in many cases differentiated pre-modern environmental history from its modern sibling. It is very easy to presume that pre-modern and modern environmental history are at odds with each other, but that is a trap to be avoided. They undoubtedly have a great deal in common, sharing many of the same questions, concerns, and methods. The modern era gets much more attention because of the implied policy considerations and the apparent immediacy of many of the issues raised. The modern era also produces much more data on which to base any discussion. In the cacophony created by concern for the present, one needs to make sure that the pre-modern history of people and the world around them is not forgotten and that the egomaniacal obsession with the here and now does not swamp what is an extremely informative and sobering antidote to the short-term considerations that dominate much discussion of the environment. One way to achieve this goal and to make the most of the examination of the pre-modern era has been to exploit existing and ongoing research in other closely related fields.

Archaeology has emerged as a critical helpmate to pre-modern environmental history. Richard Hoffmann has demonstrated, through the use of archaeology and especially archaeozoology in his study of fishing, how asking questions of archaeologists can yield a whole new body of data. One result is some redirection in the work of those paleoarchaeologists in fields like dendrochronology, paleobotany, sediment studies, and many others as they see new potential for the information they have gathered. Environmental historians bring together these findings with the written record, the confrontation enriching both sources. This kind of exploitation of multiple sources is a hallmark of pre-modern environmental history. It is a way to extract more from a highly limited base, which is in general a necessity for those working on history before Napoleon or, in the case of environmental history, before Hitler. The melding of a variety of sources gives a certain stamp to the field but also forces it to have a broader foundation than its modern counterpart.

Pre-modern environmental history is less concerned with thresholds of change, concentrating more on long-term developments and the ebb and flow of animals, plants, and the physical world than with the dangers of reaching tipping points and creating irreversible change. That is not to say that people before the modern period did not change their environments irreparably. The erosion of soils after deforestation left land barren. Richard Hoffmann has explored the irreversible changes in rivers and streams in medieval Europe caused by human action which led to the creation of new habitats for different plants and fish. The history of the era does include tipping points, but is more often typified by interaction of people and environment. The example of the Netherlands, a country where environmental history is thriving, offers a case of people acting and reacting over centuries to varied threats, some created by them. Residents of the Rhine Delta over the last millennium and a half have found novel ways to use their environment, exploring new possibilities while shifting their own strategies in dealing with a changing landscape.[7] Such a history decreases a concern for changes that cannot be stopped or redirected. Pre-modern environmental history is no less concerned with environmental footprints or relief from environmental pressure through the use of newly available resources. It does not, however, share the fear of irrevocable change. In the last century the rise in the numbers of people and the extension of their use of natural resources have made what were once matters of local concern global. Exhaustion of some physical assets or the extinction of some plant or animal from the entire planet, something that did happen in the distant past, is now not only conceivable but anticipated and has become a constant source of anxiety. In pre-modern environmental history the tendency is to explore linear effects where patterns

---

[7] One of the earliest examples of medieval environmental history, William H. TeBrake's *Medieval Frontier: Culture and Ecology in Rijnland* (College Station, Texas, 1985), started what is now a whole range of studies about management of the natural environment through the early modern and into the modern period. For a recent example of this line of research, see Milja van Tielhof and Petra J. E. M. van Dam, *Waterstaat in stedenland: Het hoogheemraadschap van Rijnland voor 1857* (Utrecht, 2006). The journal *Jaarboek voor Ecologische Geschiedenis* is a vehicle for the publication of a broad range of environmental studies. Dutch scholars have also consistently published in English in that journal and elsewhere to spread the results of their work as widely as possible. See, for example, Petra J. E. M. van Dam, "Sinking Peat Bogs: Environmental Change in Holland, 1350–1550," *Environmental History* 6 (2001), 32–45; and the essays in a special issue of *Technology and Culture* 43 (2002) devoted to technology and water resource management.

continue along general trends. For modern environmental history the concern is with nonlinear effects where moving across some threshold will cause the trend to take a dramatic new direction with no turning back.[8] Tipping the scales so far as to make restoration impossible was something rare in the pre-modern world. When there were setbacks for humans, when there were environmental disasters, or when people pressed on their environment beyond its potential, often the implications were retrenchment and restoration of earlier conditions or more simply the taking-up of a new direction in the relationship of people with the environment. The difference in the character of the nineteenth and even more so the twentieth century from more distant eras hangs over modern historical study, making the problems examined and the expectations about behaviors different. The desire to preserve, to protect some part of nature from the activities of humans which may lurk behind writing on modern environmental history, is much less an influence when considering the pre-modern past. The idea of revolutions in the relationship of humans with nature has gained limited traction among those studying the pre-modern world exactly because the changes were slow and often were a reflection of some equally impressive changes in human society.[9]

Certainly it is understood that nature was not static in the pre-modern world. The long-standing engagement with what happened to climate over the centuries is the most obvious example. In a series of books and papers starting in the mid-1960s, H. H. Lamb showed how it was possible to trace the rise and fall of temperatures and rainfall over the long term.[10] His work not only fit into the growing concern over climate change, but also formed a basis for the discussion of periods in the history of climate and specific cases where changes in weather and climate affected the economy, society, and politics.[11] In such studies

[8] J. R. McNeill, *Something New Under the Sun: An Environmental History of the Twentieth-Century World* (New York and London, 2000), pp. 3–5.

[9] See Carolyn Merchant, "The Theoretical Structure of Ecological Revolutions," *Environmental Review* 11 (1987), 265–74, for a direction urged on environmental historians.

[10] Among many works by H. H. Lamb, see *The English Climate* (London, 1964); *The Changing Climate: Selected Papers* (London, 1966); and *Climate, History, and the Modern World* (London and New York, 1982).

[11] Among many other publications in the last fifty years, see Jean M. Grove, *The Little Ice Age* (London and New York, 1988); and *Climatic Variability in Sixteenth-Century Europe and its Social Dimension*, ed. Christian Pfister, Rudolf Brázdil, and Rüdiger Glaser (Dordrecht, 1999).

of climate, nature is not a passive object but an actor and people are victims, unwitting and often uncomprehending, of the larger forces that dictate their lives. There is also concentration on measurement of changes and especially on the range of effects that can be attributed to short- and long-term variations in climate. For the pre-modern era, though, it is people who are often sources of data. Collectors and recorders of information about temperature and rainfall and other details generate localized observations that are pieced together and supplemented by economic data and what can be found in the ground and in layers of ice to make for a larger picture of climate trends.

In the minds of historians attacking the pre-modern era, people can be actors who affect their environment. The tendency is to talk more about choices made by people in dealing with the changing circumstances they faced. Perhaps it is the roots in economic history that help to drive this interest in human actions. While environmental historians are concerned with how humans deal with the constantly changing world around them, economic historians are interested in how people deal with the problem of scarcity. The questions raised by both groups are in many cases the same, as are the answers. Richard Hoffmann starts undergraduates off with an examination of production and uses of goods and that forms the basis for further examination of the pre-modern environment. Since interest is in strategies chosen by people and in the long-term shifts in both their circumstances and adjustments, pre-modern environmental historians rarely express the overt condemnation of the way people dealt with nature that can often be found in, and which often underlies, work on more recent history. That is not to say there is ignorance of error in the past or that pre-modernists do not describe and analyze damage to the environment. They are willing to draw conclusions but at the same time their concern for documentation of evolution over the long term makes them less interested in disapprobation of human action and more interested in the context of adjustment.

A corollary of this appreciation of long-term change and the scale of the human enterprise in the pre-modern world is an absence of terror regarding economic growth. By the time of the Crystal Palace Exhibition of 1851 in London there were signs that what has come to be called the Industrial Revolution was paying off in more goods and services within the reach of rapidly growing numbers of people. The unprecedented massive economic growth from then to the outbreak of the First World War in 1914, not just in Britain or in Europe but in ever more parts of the world, seemed to offer solutions to many of the problems that had up to that point always dogged human existence. Increasing production at a pace consistently faster than the growth in population was not a promise of the pre-modern world. It was unknown and not anticipated. Rapid growth was also not a threat to people or to the environment. The concern over the rising inequalities among people and in different parts of the world as a result of modern economic growth, something that has dominated the internal politics of most countries and influenced international relations over the last century and a half, was not part of the pre-modern world and so is not in the purview of those conducting research in pre-modern environmental history. That is not to say that domination, or hegemony, or slavery, or exploitation of labor or other features of the pre-modern world economy are deemed to be good or even acceptable. It is only that inequality in access to goods and services is not the pressing question that it has been for the last one hundred and fifty years.

Pre-modern environmental historians are not blind to the transformations people made in natural systems. There was over the long-term significant disruption of the ecology thanks principally but not exclusively to agricultural exploitation.[12] The consciousness of such destruction and deviation from previous conditions gives pre-modern environmental history a place alongside the modern version and the critical function of drawing thinking away from the short term and immediate. Another result is the search for cooperation among people to protect the environment when it is under siege. Common resources typically need protection if they are to survive. There have to be constraints on users in order to avoid the exhaustion if not the outright destruction of the resource. Fishers have long recognized the dangers of such overuse but people on land in the pre-modern world were also well aware of the need to regulate the use of common fields and woodlands. Cooperation to protect certain parts of the environment, a drive not simply to increase production but rather to ensure long-term exploitation, is something of continuing interest to those who study pre-modern history. There are many cases and an opportunity to see the development and evolution of such cooperation over a long period.

[12] Richard C. Hoffmann, "A Longer View: Is Industrial Metabolism Really the Problem?" in *Nature, Society and History: Long Term Dynamics of Social Metabolism*, ed. M. Fischer-Kowalski, E. Rosa, R. P. Sieferle, and B. Smetschka, as a special issue of *Innovation: The European Journal of Social Sciences* 14 (2001), 143–9.

Such examples also create a glimmer of optimism about the ability of humans to protect the environment when interests coincide.

Pre-modern environmental history offers a mild antidote to the disinterest in matters material that marks some of cultural studies. That is not to say that there is an implied attack on the study of culture. Rather there is a concern among pre-modern environmental historians for the attributes of culture, for what people extracted from their surroundings and how they used those objects from nature in defining who they were. Richard Hoffmann again offers a case in point in his work on what Europeans thought of the varied species of fish that they consumed.[13] Consumption of certain types of fish in certain places at certain times of year bestowed status and prestige. Which fish someone ate was a sign of income and of taste. In observing Lent, faithful Christians had to consume fish as their main source of animal protein. Environmental history is by definition about the material world. That has directed those studying the pre-modern period to examine consumption and the effects of that acquisition of material possessions, no matter how ephemerally, on all parts of the interrelated environment in which people lived. It is all too easy to lose sight of the importance of the material aspect of the pre-modern world which was one of shortages. That distant past did not enjoy the affluence of the modern world. The sources that pre-modern environmental historians rely on to sketch their understanding of the past are a valuable reminder for them of the critical role of the material world not only to survival, but also to position and status in a world that proved slow to change but was never completely stable.

A second corollary of the acceptance of human action and the extensive role people have long played in affecting the world around them, and the other way around, is a concern for long-term and shifting relationships between humans and their environment. Contemporary environmentalism, and with it environmental history, owes much to the recent experience in the United States of America. The history of the Great Republic is, by the standards of pre-modern history, a short one. The tendency to deal in short-term developments and so to

[13] Among a number of different articles which treat the topic in various ways, see Richard C. Hoffmann, "Medieval Fishing," in *Working with Water in Medieval Europe: Technology and Resource Use*, ed. Paolo Squatriti (Leiden, 2000), pp. 336–43; and idem, "Frontier Foods for Late Medieval Consumers: Culture, Economy, Ecology," *Environment and History* 7 (2001), 141–51.

concentrate on dramatic changes is not a luxury open to pre-modern environmental history. Rather, concentration is more often on the long-term shifts in the relations of animals to plants and the relations of both to the physical environment. It is somewhat unfair to claim that there is no sense of long-term development in American environmental history just as it is unfair to claim that cataclysm is unknown to pre-modern environmental history. The transfer of pathogens to the Western Hemisphere when Europeans made their way to what they saw as a New World, beginning in 1492, certainly qualifies as a dramatic event and one with immediate impact.[14] This event in the hands of pre-modern environmental historians has come to be placed in a larger context of the exchange between the New World and the Old of insects, animals, and plants as well as the context of the more general exchange of pathogens, flora and fauna over a millennium and more.[15] For those working on longer periods of time there are more signs of continuity than of dramatic difference; more evolution and flow of change than sharp departures. Catastrophes may make for good reading. This gives change greater importance and more immediacy but must be seen within a scale of human time, which is much more limited than geologic time. In fact, one function taken over by pre-modern environmental historians is to combat such fascination with disaster. Rather, they see a world of action and reaction, of change that can take different directions. They remind people about the nature of an environment in the past rather than the one which exists now where the forces of "biological imperialism" and unprecedented economic growth may have changed a landscape or an entire environment and thereby changed the conception of what it should be like.[16] Both nature and humans are historical actors. People and the environment interact. One case that has drawn the attention of Richard Hoffmann is the introduction of carp into central and western Europe and then the development by humans of ways to increase carp populations and control access to

[14] For a well-known summary, see Alfred W. Crosby, Jr., *The Columbian Exchange: Biological and Cultural Consequences of 1492* (Westport, Connecticut, 1972).

[15] William H. McNeill, *Plagues and Peoples* (Garden City, New York, 1976); and Alfred W. Crosby Jr., *Ecological Imperialism: The Biological Expansion of Europe, 900–1900* (Cambridge, 1986), among many others.

[16] The best example of such work which argues for the advantages of long-term study of the environment, focusing on both continuity and change, is Peregrine Horden and Nicholas Purcell, *The Corrupting Sea: A Study of Mediterranean History* (Oxford, 2000), esp. pp. 298–300.

that new food source.[17] The result through the later Middle Ages and the Renaissance was an elaborate system to generate habitat for the fish and to regulate access to what became a standard source of sustenance for many people. The exploitation of the new species meant a transformation in environments in a number of places. It also meant a new food to fit into a hierarchy of consumption and many more carp.

Pre-modern environmental history, in part because of its varied roots, leaves its doors open to a whole range of historians. There are no restrictions and those working in many different periods and places are invited to participate in the larger enterprise. Certainly there is a desire among many economic, social, and cultural historians to incorporate knowledge of environmental circumstances into their own research and increasingly to participate in the study of the environment itself. The work of Richard Hoffmann, his publications but also his organization of sessions at conferences, along with the actions of a number of others, have brought new recruits into the field of pre-modern environmental history. These are not just new entrants to the profession, raised in an era of greater ecological consciousness, but also established scholars who can, for example, find in the study of climate, changing weather, and views of medieval people about control over weather keys to understanding the problems of empires, the challenges of survival in times of crisis, and the shifting character of faith.[18]

The recruiting of new converts opens the door to a danger identified frequently by Richard Hoffmann. Practitioners in their enthusiasm may

---

[17] Richard C. Hoffmann, "Remains and Verbal Evidence of Carp (*Cyprinus carpio*) in Medieval Europe," in *Fish Exploitation in the Past: Proceedings of the 7th Meeting of the I.C.A.Z. Fish Remains Working Group*, ed. Wim Van Neer (Tervuren, 1994), pp. 139–150; idem, "Environmental Change and the Culture of Common Carp in Medieval Europe," *Guelph Ichthyology Review* 3 (1995), 57–85; idem, "*Carpes pour le duc* ...: The Operation of Fish Ponds at La Perriére-sur-Saône, Burgundy, 1338–52," *Archaeofauna* [Madrid] 4 (1995), 33–45; idem, "Carp, Cods, Connections: New Fisheries in the Medieval European Economy and Environment" in *Animals in Human Histories: The Mirror of Nature and Culture*, ed. Mary J. Henninger-Voss (Rochester, New York, 2002), pp. 3–55; and idem "Der Karpfen (*Cyprinus carpio* L.): Der lange Weg eines 'Fremdlings' in die Schweiz," in *Fisch und Fischer aus zwei Jahrtausenden: Eine fischereiwirtschaftliche Zeitreise durch die Nordwestschweiz*, ed. Heide-Marie Hüster-Plogmann (Augst, Switzerland, 2006), pp. 161–7.

[18] See, for example, William C. Jordan, *The Great Famine: Northern Europe in the Early Fourteenth Century* (Princeton, 1996); Paul Dutton, "Thunder and Hail over the Carolingian Countryside," in *Charlemagne's Mustache and other Cultural Clusters of a Dark Age* (London, 2004), pp. 169–188; and Paul Dutton, Paul A. Mayewski and Michael McCormick, "Volcanoes and the Climate Forcing of Carolingian Europe, A.D. 750 to 950," *Speculum* 82 (2007), 865–95.

ignore the development of the background in science in general and especially in the emerging field of ecology. The problem is not unique to environmental history. Many people who write about the past do so with knowledge of time and place, of language and means of communication, but lack the skills that come with training in disciplines that will help them to understand the phenomena they study. Anthropology, economics, linguistics, political science, and, to some degree, sociology have things to teach historians. The theoretical or experimental help-maidens of the historian though are all too often underutilized. This is especially marked in environmental history since so few historians come with training in biology, chemistry, or engineering that can improve appreciation of the effects of human action and the changes in flora and fauna that are the necessary outcomes of changing circumstances. In fact, historians are often driven from considering work on environmental topics because of their lack of scientific knowledge. In a number of cases they should have followed those instincts but instead have gone ahead to produce work that is of lower quality than it ought to be. It is right that scientific knowledge is a necessary asset. That context makes it possible to exercise critical skills when dealing with sources, be they descriptive, physical objects, or landscape. Richard Hoffmann has consistently urged his students and indeed any historian to gain the tools before embarking on serious work in environmental history. He has been a practitioner of what he preaches as well, developing an impressive command of archaeobiology and ichthyology. His example, along with that of many colleagues, has all but destroyed the assumption that pre-modern environmental history can be done in the absence of some knowledge of the relevant science. The trend is clearly toward a greater appreciation of what science has to offer and a more effective integration of that information into history, done typically without losing the thread of historical narrative or driving away potential readers by overreliance on highly technical information.

One topic of pre-modern environmental history, central in its early days, now appears to be fading from among concerns in the field. In 1980 when lectures on environmental history given at the University of Arkansas at Little Rock were published, the editor Lester Bilsky gave them the title *Historical Ecology: Essays on Environment and Social Change*. The goal was to aid understanding of the environmental problems of the day through the examination of similar problems from the past. The distinguished medieval historian David Herlihy contributed an essay entitled "Attitudes toward the Environment in Medieval

Society."[19] The question of what people have thought about "nature" over the long term was raised by another distinguished medievalist Lynn White Jr. in an exploration of "The Historical Roots of Our Ecologic Crisis."[20] He had argued that something inherent in Latin Christian society, dating from the Middle Ages, was at the root of the air pollution that plagued his native Los Angeles and all the other problems of environmental degradation. Herlihy was loath to accept such a singular understanding of nature as something to be subjugated and exploited and instead set forth four different attitudes toward the environment in medieval culture and by implication in subsequent centuries. One, a product of Christian eschatology, was to evaluate the present state of the world in terms of its inevitable end. A second was a fear or awe of nature. A third was that people could shape and change the environment to improve the human condition now and in the after-life. The fourth was to see the natural world as a place for refreshment, renewal, and even recreation. Each had greater traction at different times but all were present at almost all times. Although Herlihy may have broadened the scope of the discussion, this did not overcome the suspicion made explicit by White that some essential feature of Western thinking created the precondition for an assault on the natural world. The concern then for ideas about nature became a way to approach environmental history over the long term and worldwide.[21] The musings of intellectuals about what attitudes toward labor might imply for relationships to the natural world followed an established discussion of concepts of work among social historians but brought that discussion into the realm of environmental history.[22] While questions about the roots of an aggressive approach to nature, one which would set Western industrial societies and the people who live in them apart from other peoples and cultures, have not disappeared, interest in such

---

[19] David J. Herlihy, "Attitudes toward the Environment in Medieval Society," in *Historical Ecology: Essays on Environment and Social Change*, ed. Lester J. Bilsky (Port Washington, New York, 1980), pp. 100–116.

[20] Lynn White, Jr., "The Historical Roots of Our Ecologic Crisis," *Science* (10 March 1967), reprinted in idem, *Dynamo and Virgin Reconsidered: Essays in the Dynamism of Western Culture* (Cambridge, Massachusetts, 1968), pp. 75–94.

[21] Gilbert F. Lafreniere, *The Decline of Nature: Environmental History and the Western Worldview* (Bethesda, Maryland, 2007), is a recent and excellent example.

[22] Jacques Le Goff, *Time, Work and Culture in the Middle Ages*, trans. Arthur Goldhammer (Chicago, 1980), esp. pp. 58–121; and George Ovitt, *The Restoration of Perfection: Labor and Technology in Medieval Culture* (New Brunswick, New Jersey, 1987).

distinctions has waned and with good cause. As more has become known about the complexity of European society and its daughters in Australasia and the New World and about the other varied religions and societies that have lived on the globe, the sharp distinctions that White suggested have become blurred. As Herlihy already noted in 1980, not everyone, even in medieval Europe, thought the same way about how to approach the environment. Ideas about nature no longer draw the same interest as before. Attention has turned instead to the place of people of all kinds and backgrounds in nature. The record of human action, now ever fuller thanks to rapidly expanding research on a range of topics and in various places around the world, is creating a body of information formerly unknown. That greater knowledge of material action is proving more productive in pre-modern environmental history than discussion, largely generated by intellectual historians, about the thinking of the more verbal members of western society.

A second essay in the same 1980 collection by Charles Bowlus dealt with "Ecological Crises in Fourteenth-Century Europe." The mid-century demographic disaster attributed to an environmental cause, specifically the reintroduction of the plague bacillus into western and central Europe, was the focus of his essay. Whether the Black Death was in fact really a bubonic plague has generated extensive discussion in the intervening years and offered a productive topic for pre-modern environmental historians.[23] Even more telling in that contribution though was the focus on the concept of crisis, a central feature of the recent American Historical Review Conversation in which Richard Hoffmann was a principal player.[24] The idea of a 'crisis' that requires immediate action grew out of the political rhetoric of the Cold War in the United States and was already overused by the time environmentalists began to exploit it to describe the need for direct and immediate action in stemming the tide of destruction they saw around them. The participants in the AHR Roundtable, though they may have employed the word, were in general skeptical of the usefulness of what

[23] Charles Bowlus, "Ecological Crises in Fourteenth-Century Europe," in *Historical Ecology*, pp. 86–99; Samuel K. Cohn, Jr., "The Black Death: End of a Paradigm," *American Historical Review* 107 (2002), 703–38; and idem, *The Black Death Transformed: Disease and Culture in Early Renaissance Europe* (New York, 2003).
[24] See n. 1, above.

has become a rather imprecise concept. The work of pre-modern environmental historians has not dimmed the concern for contemporary problems but it has, by taking a long-term view and exploring varied regimes around the world, eased anxiety and made more transparent the need for sustained concern for and involvement in the nurturing of nature and seeing the interaction of people and their environments in many dimensions and all at once.

While the concept of a crisis may not be useful, environmental disasters have for historians proven helpful in understanding societies. The sudden stress created by the collapse of crops or destruction from weather or earthquakes can, historians have found, produce valuable knowledge about political and social structures and what people thought of themselves and the way their local societies were organized.[25] Certainly such work qualifies as environmental history but it is not about the environment. Rather it exploits effectively knowledge of the environment as a way to explore even larger issues. The principle is the same for long-term studies of human interaction with the rest of the world, but the concentration is not on a specific disaster in a particular context but rather on the ebb and flow over time of the ways in which people deal with circumstances that they have a role in changing.

Pre-modern environmental history offers new dimensions to other already well-established fields of history. In a sense this is remuneration to those aspects of historical study that were the roots of the subfield. Environmental history cannot now be characterized as a minor subdivision of the discipline as it could rightly have been in 1995.[26] It is instead a necessary part of many types of scholarly works about the human past. Studies of the pre-modern economy, politics, culture, food preparation and consumption, and even military history have benefited from the results of what is now known about the relationship of people to nature. Often, almost of necessity, historians in various fields in recent years have come to include recognition of the role of the environment in shaping outcomes. Writers these days feel compelled, and overtly so, to take the environment into account. An emerging product of that compulsion is new kinds of synthesis. The many smaller studies

---

[25] A recent example of such "incident analysis," where an environmental disaster is the trigger, is Charles F. Walker, *Shaky Colonialism: The 1746 Earthquake-Tsunami in Lima, Peru, and its Long Aftermath* (Durham, North Carolina, 2008).

[26] Crosby, "Past and Present of Environmental History," p. 1188.

of local developments, short-term changes, or specific interactions between people and their surroundings has built up enough of a foundation for the production of general surveys of pre-modern environmental history. The task is not an easy one. Authors have in some cases chosen a common theme to join what are many disparate case studies.[27] There is at least one attempt to bring a broad range of research together, an effort in which Richard Hoffmann took part.[28] As a result of the steady stream of specific studies, there will be more general histories that draw on that growing body of knowledge. It will become possible to take one thread or strand or issue in the history of the environment and follow it tenaciously over time and across a broad geographical area. There will be a concentration on nuance and complexity, made possible because so much work has now been done and made necessary because now all the easy answers to questions about the way people deal with the world around them have been proven wrong.

Pre-modern environmental history is now less self-conscious than it was. It is less concerned with ideas about nature. It is more about science and the use of science to inform analysis of the past interactions of people and their environments. It is about periods and places formerly neglected. It is about seeing places holistically as part of complete landscapes of land, people, animals, and plants.[29] It is about dealing with past societies on their own grounds and in their own terms rather than imposing twenty-first century criteria on them. It is less concerned with reflecting on what it is, as in this discussion, and more concerned with the formation, through practice, of a vibrant field. Richard Hoffmann has been one of those who has broken the trail for aspiring environmental historians and played a role in the upbringing of this burgeoning beast. He has directed his effort at pursuing fish, specifically in finding the connections among the different treatment by people of the many animals that fit into that broad category over a long period of time. The by-products of this large study have already been numerous from specific studies of fishing practices,

---

[27] For an excellent example of the writing of global environmental history, see the very ambitious study by John F. Richards, *The Unending Frontier: An Environmental History of the Early Modern World* (Berkeley, 2003).

[28] Richard C. Hoffmann, "Medieval Christendom in God's Creation: Environmental Continuities, Coevolutions, and Changes," in *Northern Europe: An Environmental History*, ed. Tamara L. Whited et al. (Santa Barbara, Denver and Oxford, 2005), pp. 45–72.

[29] Horden and Purcell, *The Corrupting Sea*, pp. 176–8.

the administration of fishing rights, the relationship between environmental changes and fish populations, the character of equipment used in fishing, the emergence of recreational angling and what contemporaries said about how to fish. His work has also influenced many others to take up the examination of fishing history.[30] The approach Richard Hoffmann has adopted has produced valuable spin-offs both in what he does and in what others do. His work has reflected the more general development of pre-modern environmental history as it has evolved in his years as a practicing historian. His recommendations to others in person and in print have helped to send pre-modern environmental history in its present positive direction. He has enriched the field both by increasing knowledge of what is a multivalent and diverse past and by shaping the field, one which is robust and enjoys a bright future.

## *Bibliography*

"AHR Conversation: Environmental Historians and Environmental Crisis," *American Historical Review* 113 (2008), 1431–65.

*Beyond the Catch: Fisheries of the North Atlantic, the North Sea and the Baltic, 900–1850*, ed. Louis Sicking and Darlene Abreu-Ferreira (Leiden, 2008).

Bowlus, Charles. "Ecological Crises in Fourteenth-Century Europe," in *Historical Ecology: Essays on Environment and Social Change*, ed. Lester J. Bilsky (Port Washington, New York, 1980), pp. 86–99.

*Climatic Variability in Sixteenth-Century Europe and its Social Dimension*, ed. Christian Pfister, Rudolf Brázdil, and Rüdiger Glaser (Dordrecht, 1999).

Cohn, Jr., Samuel K. "The Black Death: End of a Paradigm," *American Historical Review* 107 (2002), 703–38.

——. *The Black Death Transformed: Disease and Culture in Early Renaissance Europe* (New York, 2003).

Crosby, Jr., Alfred W. *The Columbian Exchange: Biological and Cultural Consequences of 1492* (Westport, Connecticut, 1972).

——. *Ecological Imperialism: The Biological Expansion of Europe, 900–1900* (Cambridge, 1986).

——. "The Past and Present of Environmental History," *American Historical Review* 100 (1995), 1177–89.

Dutton, Paul. "Thunder and Hail over the Carolingian Countryside," in *Charlemagne's Mustache and other Cultural Clusters of a Dark Age* (London, 2004), pp. 169–88.

Dutton, Paul, Paul A. Mayewski and Michael McCormick, "Volcanoes and the Climate Forcing of Carolingian Europe, A.D. 750 to 950," *Speculum* 82 (2007), 865–95.

Grove, Jean M. *The Little Ice Age* (London and New York, 1988).

---

[30] The best recent example is *Beyond the Catch: Fisheries of the North Atlantic, the North Sea and the Baltic, 900–1850*, ed. Louis Sicking and Darlene Abreu-Ferreira (Leiden, 2008).

Herlihy, David J. "Attitudes toward the Environment in Medieval Society," in *Historical Ecology: Essays on Environment and Social Change*, ed. Lester J. Bilsky (Port Washington, New York, 1980), pp. 100–16.
Hoffmann, Richard C. "Warfare, Weather, and a Rural Economy: The Duchy of Wrocław in the Mid-Fifteenth Century," *Viator* 4 (1973), 273–305.
——. *Land, Liberties, and Lordship in a Late Medieval Countryside: Agrarian Structures and Change in the Duchy of Wrocław* (Philadelphia, 1989).
——. "Remains and Verbal Evidence of Carp (*Cyprinus carpio*) in Medieval Europe," in *Fish Exploitation in the Past: Proceedings of the 7th Meeting of the I.C.A.Z. Fish Remains Working Group*, ed. Wim Van Neer (Tervuren, 1994), pp. 139–50.
——. "Environmental Change and the Culture of Common Carp in Medieval Europe," *Guelph Ichthyology Review* 3 (1995), 57–85.
——. "*Carpes pour le duc* ... The Operation of Fish Ponds at La Perriére-sur-Saône, Burgundy, 1338–52," *Archaeofauna* [Madrid] 4 (1995), 33–45.
——. "Medieval Fishing," in *Working with Water in Medieval Europe: Technology and Resource Use*, ed. Paolo Squatriti (Leiden, 2000), pp. 331–93.
——. "Frontier Foods for Late Medieval Consumers: Culture, Economy, Ecology," *Environment and History* 7 (2001), 131–67.
——. "A Longer View: Is Industrial Metabolism Really the Problem?" in *Nature, Society and History: Long Term Dynamics of Social Metabolism*, ed. M. Fischer-Kowalski, E. Rosa, R. P. Sieferle, and B. Smetschka, as a special issue of *Innovation: The European Journal of Social Sciences* 14 (2001), 143–55.
——. "Carp, Cods, Connections: New Fisheries in the Medieval European Economy and Environment," in *Animals in Human Histories: The Mirror of Nature and Culture*, ed. Mary J. Henninger-Voss (Rochester, New York, 2002), pp. 3–55.
——. "Medieval Christendom in God's Creation: Environmental Continuities, Coevolutions, and Changes," in *Northern Europe: An Environmental History*, ed. Tamara L. Whited et al. (Santa Barbara, Denver and Oxford, 2005), pp. 45–72.
——. "Der Karpfen (*Cyprinus carpio* L.): Der lange Weg eines 'Fremdlings' in die Schweiz," in *Fisch und Fischer aus zwei Jahrtausenden: Eine fischereiwirtschaftliche Zeitreise durch die Nordwestschweiz*, ed. Heide-Marie Hüster-Plogmann (Augst, Switzerland, 2006), pp. 161–7.
Horden, Peregrine and Nicholas Purcell, *The Corrupting Sea: A Study of Mediterranean History* (Oxford, 2000).
Jordan, William C. *The Great Famine: Northern Europe in the Early Fourteenth Century* (Princeton, 1996).
Lafreniere, Gilbert F. *The Decline of Nature: Environmental History and the Western Worldview* (Bethesda, Maryland, 2007).
Lamb, H. H. *The English Climate* (London, 1964).
——. *The Changing Climate: Selected Papers* (London, 1966).
——. *Climate, History, and the Modern World* (London and New York, 1982).
Laslett, Peter. *The World We Have Lost* (London, 1965).
Le Goff, Jacques. *Time, Work and Culture in the Middle Ages*, trans. Arthur Goldhammer (Chicago, 1980).
McNeill, J. R. *Something New Under the Sun: An Environmental History of the Twentieth-Century World* (New York and London, 2000).
——. "Observations on the Nature and Culture of Environmental History," *History and Theory* 42 (2003), 5–43.
McNeill, William H. *Plagues and Peoples* (Garden City, New York, 1976).
Merchant, Carolyn. "The Theoretical Structure of Ecological Revolutions," *Environmental Review* 11 (1987), 265–74.
Ovitt, George. *The Restoration of Perfection: Labor and Technology in Medieval Culture* (New Brunswick, New Jersey, 1987).

Pfister, Christian. "Das '1950er Syndrom': Die umweltgeschichtliche Epochenschwelle zwischen Industriegesellschaft und Konsumgesellschaft," in *Das 1950er Syndrom: Der Weg in die Konsumgesellschaft*, ed. Christian Pfister (Bern, 1995), pp. 51–95.

Radkau, Joachim. *Nature and Power: A Global History of the Environment*, trans. Thomas Dunlap (New York, 2008).

Richard C. Hoffmann, "Medieval Christendom in God's Creation: Environmental Continuities, Coevolutions, and Changes," in *Northern Europe: An Environmental History*, ed. Tamara L. Whited et al. (Santa Barbara, Denver and Oxford, 2005), pp. 45–72.

Richards, John F. *The Unending Frontier: An Environmental History of the Early Modern World* (Berkeley, 2003).

TeBrake, William H. *Medieval Frontier: Culture and Ecology in Rijnland* (College Station, Texas, 1985).

Van Dam, Petra J. E. M. "Sinking Peat Bogs: Environmental Change in Holland, 1350–1550," *Environmental History* 6 (2001), 32–45.

Van Tielhof, Milja and Petra J. E. M. van Dam, *Waterstaat in stedenland: Het hoogheemraadschap van Rijnland voor 1857* (Utrecht, 2006).

Walker, Charles F. *Shaky Colonialism: The 1746 Earthquake-Tsunami in Lima, Peru, and its Long Aftermath* (Durham, North Carolina, 2008).

White, Jr., Lynn. "The Historical Roots of Our Ecologic Crisis," *Science* (10 March 1967), reprinted in idem, *Dynamo and Virgin Reconsidered: Essays in the Dynamism of Western Culture* (Cambridge, Massachusetts, 1968), pp. 75–94.

# PART ONE

# PREMODERN PEOPLE AND THE NATURAL WORLD

# TREES, NUTS, AND WOODS AT THE END OF THE FIRST MILLENNIUM: A CASE FROM THE AMALFI COAST

Paolo Squatriti

The cartulary of the Benedictine convent of S. Lorenzo al Piano, founded just outside Amalfi in the late 900s, is commonly known by the name of the Enlightenment lawyer from Angri who purchased it around 1780 for use in his practice, as an arsenal of legal precedent. The "Perris Codex" is comprised of almost six hundred documents and contains several early medieval charters, though the bulk of its content is late medieval, of deeper interest to the copyist who created the extant version.[1] Among the documents from the tenth and early eleventh centuries, several make reference to a characteristic land use of southern Campania in those times, the chestnut grove. This essay analyzes only one charter in any detail, but attempts to build on it a wider portrait of medieval chestnut cultivation around Amalfi. *Castanea sativa*, the European chestnut, remains a significant presence in Campania's highlands today, yet the early charters suggest that along the Amalfi peninsula, in the Lattari mountains looming above the maritime zone and even in some low valleys there, just as further south and east around Salerno and the Picentini mountains, and in the vicinity of Avellino further inland, this tree was once more important than it became in the twentieth century.[2] In early

[1] In 1943 SS. Trinità's original charters in the Archivio di Stato at Naples were destroyed. The "Perris Codex" was still in private hands then, and survived. It seems to be a fifteenth-century compilation: Jole Mazzoleni and Renata Orefice, ed., *Il codice Perris* (Amalfi, 1985), pp. xi–xii; and Antonio Allocati, "Il cartulario amalfitano detto comunemente 'Codice Perris,' e la sua edizione," in *Convegno internazionale 14–16 giugno 1973. Amalfi nel medioevo* (Salerno, 1977), pp. 361–65.

[2] Though most of them reflect ecologies different from Amalfi's, the Cava and Montevergine charters confirm the impression left by the "Codice Perris," and the *Codice Diplomatico Amalfitano*, ed. Riccardo Filangieri di Candida (Naples, 1917). For a synoptic view, see ch. 3 in my *Dark Ages and Old Chestnuts in Europe* (forthcoming). On the decline of chestnut cultivation since the 1800s, see Massimo Becchi, *Discorso sul castagno* (Reggio Emilia, 1996), p. 27; and Marco Conedera et al., "Competition and Dynamics in Abandoned Chestnut Orchards in Southern Switzerland," *Ecologia Mediterranea* 26 (2000), 101–12.

medieval Campania, it appears, the botanical properties of *Castanea sativa* suited prevailing economic and social conditions enough to create a distinctive woodland landscape different from what had come before and what was to come. In this dynamic landscape chestnuts did not occupy marginal terrain or preoccupy impoverished categories of people. On the contrary, chestnut cultivation in postclassical Campania was integral to the ebullient commercial activities that made the central Tyrrhenian so unlike the Pirennian stereotype of Dark Age Mediterranean stasis and autarchy. Thus the chestnuts were more than stolid occupants of the hillsides neglected by humans. They were agents in the environmental transformation of the early Middle Ages.

With the charters allotting so much space to what they called "castanieta," it is quite logical that scholars have devoted some attention to chestnut cultivation in early medieval Italy and Campania. Conceptualizations of the place of chestnuts in the early medieval Italian economy and agriculture have, however, been varied. Because of their evaluation of Mediterranean economic stability between the Neolithic and Industrial revolutions, for example, Horden and Purcell opine that from ancient times *Castanea sativa* "provided an alternative economy" to grains, a fall-back food when sown crops failed, or a swine food in normal years, a means of turning less favored environments into cash through the sale of pork fattened on the nuts.[3] In this version of chestnut history (or "castaneology"), nothing much changed in the early Middle Ages. More botanically informed accounts, like that of Grove and Rackham, note instead that chestnuts did increase in importance in the postclassical centuries, but also remark that this transformation is "ill-documented."[4] The recent account of Tuscan chestnut cultivation by Quirós Castillo refutes both of these positions by demonstrating that chestnuts not only rose to prominence in the Dark Ages as settlement shifted to new hillside locations, but also that archaeobotanical and documentary evidence for this rise exists.[5]

---

[3] Peregrine Horden and Nicholas Purcell, *The Corrupting Sea: A Study in Mediterranean History* (Oxford, 2000), pp. 117, 199 and 203. Yet early medieval pigs ate acorns, and people kept the chestnuts for themselves (an example from AD 1001 is in Placido Tropeano, ed., *Codice Diplomatico Verginiano*, 13 vols. [Montevergine, 1977–2000], vol. 1, p. 74).

[4] A. T. Grove and Oliver Rackham, *The Nature of Mediterranean Europe* (New Haven, 2001), p. 177.

[5] Juan Antonio Quirós Castillo, "Cambios y transformaciones en el paysaje del Appennino toscano entre la Antigüedad Tardía y la Edad Media," *Archeologia medievale* 25 (1998), 180–5.

Variety of interpretation also characterizes the older "castaneological" literature. Montanari's classic description of the compenetration of cultivated and uncultivated sectors of the landscape in the early medieval economy had identified the chestnut as a special case, or a type of land use of peculiar relevance in postclassical times.[6] To Montanari, the chestnut-filled portions of the landscape were ambiguous and liminal, neither wild nor agrarian. They were precious most of all for the peasant cultivators who ate their fruit, but were useful also to elites who could take rents and tributes of various kinds from the trees. To some extent, Montanari's view of the evolution of chestnut woodlands in the postclassical centuries mirrored that offered a few years earlier by Pierre Toubert. Toubert's great book on the structures of medieval Latium offered one of the first and most insightful reconstructions of the place of *Castanea sativa* in a medieval Italian landscape and launched medieval "castaneology" on its course. For Toubert, chestnuts were barely tolerated by the lords who reconfigured rural settlement in the Sabine hills during the tenth century and who exploited the new agrarian landscape thereafter. To the extent that chestnut groves existed in Toubert's Sabina, they were relics of cleared woodlands, kept on the remotest edges of the productive agricultural space. In the Latial hills around 1000, chestnuts were a transitional land-use, between the wild woodland of the Dark Ages and the cleared, fully agricultural high medieval future. In this conceptualization, the chestnuts were a way for wily Sabine peasants to keep some marginal land more or less productive without great investment of labor. Demographic pressure and the more market-oriented seigneurial agriculture that prevailed in the high Middle Ages ineluctably cleared chestnut woods, replacing them with olive groves and especially with vineyards in the twelfth and thirteenth centuries.[7] Toubert's analysis of twelfth-century Campanian agriculture also stressed the innovations of the high medieval period and the unprecedented reliance on chestnuts as a way of extracting products from marginal, hilly land in a time of increased

---

[6] Massimo Montanari, *L'alimentazione contadina nell'alto medioevo* (Naples, 1979), pp. 38–43 and 296–301 (see also his *Campagne medievali* [Turin, 1984], pp. 157–9). Giovanni Cherubini, "La 'civiltà' del castagno in Italia alla fine del medioevo," *Archeologia medievale* 8 (1981), 247–80 focused on later times but was a milestone in medieval "castaneology" and makes many useful considerations.

[7] Pierre Toubert, *Les structures du Latium médiévale* (Rome, 1973), pp. 177–9, 190–2 and 345–7.

demand for food.[8] Toubert's view was in essence preserved in the overview Giovanni Vitolo offered of Campanian chestnut cultivation, with a rigid distinction between an early medieval period of low productivity and subsistence gathering of "wild" fruit, and a much more commercial and scientific cultivation of chestnut trees after 1000.[9]

The Amalfitan case presented here is an expansion of these earlier analyses of how chestnuts fit into the postclassical landscape. It differs by suggesting that the biological properties of chestnut trees were important historical agents and can help illuminate the transformation of the countryside in Campania. In particular, *Castanea sativa*'s productivity and its extraordinary amenability to human care proved to be a winning combination in the early Middle Ages. The case also suggests that some distinctions commonly made between early and late medieval land use bear reconsideration. Though we are accustomed to distinguish between an earlier medieval agrarian production for subsistence, and its special crops, and a later medieval agrarian production in response to market demand, with its own preferred plants, on the Amalfi coast chestnut trees managed to fill both roles.[10] Thus I suggest that chestnut cultivation in this part of Campania began at the beginning of the Middle Ages, as a solution to labor scarcity and new settlement patterns. It continued to flourish even after the demographic and commercial revival of the late first millennium, finding new outlets and purposes in the increasingly market-oriented Campanian economy around 1000. One of the many insights offered by Richard Hoffmann's study of medieval European environmental relationships is that "Europe's medieval demographic experience … is central to its environmental history."[11] By applying this idea to an early medieval case from near Amalfi recorded in the "Perris Codex," I intend to show that *Castanea sativa* was neither a marginal nor a transitional presence on

[8] Pierre Toubert, "Paysages ruraux et techniques de production en Italie méridionale dans la seconde moitié du XIIe siècle," in *Potere, società e popolo nell'età dei due Guglielmi* (Bari, 1981), pp. 208–10.

[9] Giovanni Vitolo, "Il castagno nell'economia della Campania medievale," *Rassegna storica salernitana* 12 (1989), 21–34. Earlier, Mario del Treppo, "Una città del Mezzogiorno nei secoli IX–XIV," in *Convegno internazionale*, pp. 44–5 and 53 had raised the same issues.

[10] Chris Wickham, "Agricultura, ambiente e sviluppo economico nella storia europea: il problema dell'alto medioevo," in *Agricultura ambiente e sviluppo nella storia europea*, ed. Luciano Segre (Milan, 1993), p. 158.

[11] See his "Medieval Christendom in God's Creation," in *Northern Europe: An Environmental History*, ed. Tamara Whited et al. (Santa Barbara, Denver and Oxford, 2005), p. 47.

the hillsides of Campania, but interacted with people dynamically, always establishing an economic and environmental relationship with them in harmony with local demographic levels. In a nutshell, what I will propose is that the lighter population densities that seem to have prevailed in Europe, in Italy, and in Campania after the sixth century, formed ideal conditions for the propagation of *Castanea sativa*.[12] Once this propagation had occurred, *Castanea sativa* became rooted in the slopes of the Lattari and Picentini, firmly enough that it was relevant in the more densely populated high Middle Ages and indeed firmly enough that the tree remains a meaningful land use there today.[13]

The classic studies of medieval demography are decades old, and revision of them proceeds unevenly on account of the intractable evidence.[14] Yet early medievalists accept that there were far fewer people in postclassical Europe than there had been at the height of Rome's imperial hegemony. Specialists in the history of the Italian peninsula are no different in this regard from scholars of the other former provinces of the Roman empire.[15] If there was a Dark Age demographic collapse, and only slow, tentative demographic recovery in the last two centuries of the first millennium AD, we should expect the new demographic reality to have had environmental effects (as well as causes).

---

12 Linking chestnut crops to subsistence, Cherubini, "La 'civiltà,'" pp. 268–9 argues the opposite, namely that higher population levels led to more chestnut cultivation in late medieval times. I agree with Ariane Bruneton-Governatori, "Alimentation et idéologie," *Annales: Economies, société, civilisations* 39 (1984), 1181, who thinks new chestnut plantations never derived from subsistence demand because of the length of time required for the trees to become productive.

13 Landscape scholars often note the persistence of land uses after their original purpose has subsided: Tom Williamson, *Shaping Medieval Landscapes* (London, 2003), p. 193. On twentieth-century land use, Touring Club Italiano, *Carta dell'utlizzazione del suolo d'Italia 1:200000* (Milan, 1960), foglio 16.

14 Josiah Russell's studies, based on Karl Beloch's *Bevölkerungsgeschichte Italiens*, 3 vols. (Berlin, 1937–61), need revision. For an orientation on how this might look, see Jean Pierre Devroey, *Économie rurale et société dans l'Europe franque*, 2 vols. (Paris, 2003–05), vol. 1, pp. 42–8. Fabio Giovannini, *Natalità, mortaliè e demografia dell'Italia medievale* (Oxford, 2001); and Irene Barbiera and Gianpiero Dalla Zuanna, "Le dinamiche della popolazione nell'Italia medievale," *Archeologia medievale* 34 (2007), 19–42, attempt for Italy the kind of "qualitative" demography Devroey advocates. Elio LoCascio and Paolo Malanima, "Cycles and Stability: Italian Population Before the Demographic Transition," *Rivista di storia economica* 21 (2005), 197–232 offer a new take on traditional quantitative analyses. The orthodox view: Michel Rouche, "Le haut moyen âge," pp. 133–67 and Giuliano Pinto and Eugenio Sonnino, "L'Italie," pp. 485–8, both in *Histoire des populations de l'Europe*, ed. Jean-Pierre Bardet and Jacques Dupâcquier (Paris, 1997).

15 For instance, Neil Christie, *From Constantine to Charlemagne: An Archaeology of Italy, AD 300–800* (Aldershot, 2006), pp. 57–61, 249–51, 260–3 and 500–4.

In fact, early medievalists often propose that the Roman landscape was drastically altered by the relative absence of people and that natural woodland reoccupied many riparian and highland areas reduced to treelessness by Roman agriculture.[16] In effect, the Italian peninsula after the Gothic Wars in the mid-sixth century traded people for plants. The change had repercussions on mentalities and economies, both of which became more sylvan than they had been in Roman times.[17]

Italy's early medieval woodland is not heavily frequented nowadays, but scholars agree that the wooded areas were anything but abandoned in the postclassical centuries. Aside from the people who gathered fuel and food in them, and those who hunted the mammals that lived there, the greatest utilization of the woodland was by pastoralists. Sheep and pigs, and their keepers, filled the treed landscape, to such an extent that (in one of the scholarly tropes about Italy's woods in the early Middle Ages) the standard measurement of woods' extent was by the number of pigs that found pasture in them. Woods like these were definitely not desolate wastelands. Instead they were fully integrated into the economic space of people, on par with cultivated fields and vineyards.[18]

The chestnut orchards dealt with here are therefore one of many uses of the woodland that the sparser populations of postclassical Italy devised. Still, Campania's chestnut woods differed from other woodlands in several ways. Perhaps the most significant difference lay in the amount of work that people lavished upon them: the "castanieta" of so many charters from Amalfi, Cava, and Montevergine were a human landscape, as well as a natural one, or at least mixed these two qualities in dosages unlike other woodlands. If they were not quite like a fruit orchard or a vineyard they were also not like the spontaneous woods of oaks, ash, or beech that hid behind what the notaries called "silva." The cultivation of chestnuts is a nice illustration of the flexibility of early medieval land use

---

[16] Devroey, *Économie rurale*, pp. 27–34; Chris Wickham, "European Forests in the Early Middle Ages," *Settimane di studio del centro italiano di studi sull'alto medioevo* 37 (Spoleto, 1990), pp. 499–501, 530–35; and Romualdo Trifone, *Storia del diritto forestale in Italia* (Florence, 1957), pp. 20–9.

[17] Massimo Montanari, "Dalla tardoantichità all'alto medioevo," in *Storia dell'alimentazione*, ed. Jean-Louis Flandrin and Massimo Montanari (Bari, 1997), pp. 215–23.

[18] Wickham, "European Forests," pp. 479–545 offers a good synthesis. An antidote against suspiciously Edenic visions of early medieval land use comes from the selective deforestation chronicled by J. John Lowe et al., "Stratigrafia pollinica olocenica e storia delle risorse boschive dell'Appennino settentrionale," *Rivista geografica italiana* 102 (1995), 267–95.

and the inappropriateness of applying to it rigid classical agronomical and legal categories like "ager," "saltus," and "silva," or even the more modern historiographical distinction between "incolto" and "coltivato."

For *Castanea sativa* does not propagate itself at all easily without human assistance, not least because of its bulky seed, incapable of wafting very far from the mother tree. The clumsy method of reproduction that people call the chestnut is particularly detrimental to a plant that requires abundant sunshine to grow well: under the dense canopy of mature chestnut trees very few seeds manage to grow into saplings. Compounding the reproductive challenge, the chestnut tree tends to develop its leaves late in the Mediterranean season, sometimes as late as mid-May, so any nuts that managed to roll down hills or get a lift with an animal and thus escape the shade of the mother tree would suffer from the competition for light by more precocious and vigorous growers the following spring. Furthermore, chestnuts must cross-pollinate in order to be fertile, so while a grown tree makes it difficult for its offspring to grow nearby it also needs neighbors of its same species in order to reproduce itself. This set of conditions explains why chestnut woods do not exist "in nature" in the Italian peninsula, and therefore in Campania.[19]

Ironically, it was the European chestnut's distinctive botanical characteristics that gave it a leg up in the competitive arboreal world of Dark Age Campania. *Castanea sativa* formed an alliance with people and through this alliance became one of the most prominent species in those woodlands people frequented and wrote about. For the charters leave little doubt that by the end of the first millennium cultivators in southern Campania inhabited woodlands in which the chestnut had a primary place.[20] This was because, duly encouraged, chestnuts offer their friends bountiful benefits. Amongst these the nut figures prominently, for it is sweet and highly nutritious. Although the charters mention some use of chestnut wood, and *Castanea sativa* has prodigious regenerative capacities that make it far and away the best producer of wood among Italian species of tree, it was the nuts that most interested the landlords and cultivators who compiled the Campanian charters from before 1000.

---

[19] The arboricultural literature is vast. For an orientation see Luigi Fenaroli, *Il castagno* (Rome, 1945); and Jean-Robert Pitte, *Terres de castanide* (Paris, 1986), pp. 8–37.

[20] Vitolo, "Il castagno," p. 22 tabulates charter references to suggest chestnuts ranked third among Campanian land uses.

This was a considerable change from the preferences of ancient agronomists, and perhaps peasants too. Some Roman writers (and presumably their audiences) knew of *Castanea sativa*, but to the agronomists it was a tree whose sole interest lay in the poles it produced for viticulture, or at most in the feed it generated for pigs, while to poets like Virgil, Ovid, and Martial it represented crude and backwards people, especially shepherds, their occupations, and their scavenging ways. Pliny, who described the tree accurately in his *Natural History* (15.92), seems to have thought the chestnut was on earth to grow stout posts that could become fencing, and that for the most part the nuts were indigestible, except to swine.[21] At the height of Rome's empire, when that other protagonist of Campanian environmental history, Mount Vesuvius, erupted, in an event Pliny witnessed, Campanian houses contained virtually no chestnut wood, and indeed such wood is conspicuous for its absence in Roman shipwrecks and excavated sites.[22] A single house at Pompeii, one at Herculaneum, and the villa at Oplontis contained very small quantities of chestnuts when the volcano erupted.[23] Moreover chestnut pollens, sometimes used to chronicle the expansion of the tree's cultivation in Roman provinces, are virtually absent in southern Campania, though of course this datum is liable to change with new, more refined excavations. Overall it seems that the chestnut woods in the charters from the ninth and tenth centuries were an innovation, a major landscape change originating in the preceding centuries whose charters do not survive. Likewise new was the assiduous attention that landlords and cultivators gave to the nuts themselves, evidently an important Dark Age food in the region.

The edible nut of *Castanea sativa* appears with predictable rhythms. Each autumn, a mature chestnut tree produces many nuts, each season

---

[21] Most ancient chestnut lore derived from Theophrastus, a student of Aristotle. Pitte, *Terres de castanide*, pp. 51–77 reviews the literary evidence. Always enlightening is Russell Meiggs, *Trees and Timber in the Ancient World* (Oxford, 1982), pp. 267–70.

[22] Paola Pugsley, *Roman Domestic Wood* (Oxford, 2003), p. 153 found a single box-top of chestnut wood in her survey of the western Empire's wood use. Equally measly results come from synthetic works on Pompeii (*Homo Faber*, ed. Jürgen Renn and Giovanni Castagnetti [Rome, 2003]) and Herculaneum (Stephanus Mols, *Wooden Furniture in Herculaneum* [Amsterdam, 1999]). Meiggs, *Trees and Timber*, pp. 237–47 and 365–8 suggests chestnut was little used in building houses or ships. Wickham, "European Forests," p. 536 assumed chestnut woods were normal in Roman Europe.

[23] As the eruption occurred in late August, the nuts must be from the A.D. 78 crop: Wilhelmina Jashemski et al., "Catalogue of Plants," in *The Natural History of Pompeii*, ed. Wilhelmina Jashemski and Frederick Meyer (Cambridge, 2002), p. 97.

of course differing somewhat from the next, abundant years succeeding upon years of more modest production, though without the marked fluctuations that characterize grains. On average, a mature tree in Italy produces fifty kilos of chestnuts, though in favorable years when the weather is warm and still during flowering, a tree that is younger than 150 or so years in a favorable soil can bear some two hundred kilos of chestnuts, or 70,000 calories worth.[24] Most nuts ripen in October and November (but in modern Campania some gathering extends into December). A tree's productive life begins in earnest during its second decade of life, and in its fourth decade a tree reaches its full productive potential. As chestnut trees live enormously long lives, and indeed live longer than any other Italian tree save the olive, properly cared for they become a multigenerational and sustainable provider of food.[25] Proper care is not negligible and a young grove's maintenance can be quite laborious. In areas, however, where *Castanea sativa*'s fairly rigid requirements for moisture and acidic soil are met, mature chestnuts flourish with relatively little labor. This is truest when chestnut trees are compared with some other sources of food, especially with sown crops, but also when compared to that other favored Mediterranean arboricultural choice, the vine. Peasants who did the work will have appreciated this. Some calculations done on the basis of early modern Piedmontese evidence allow the conclusion to be drawn that chestnuts were vastly superior to grains in terms of their productivity in relation to the work expended on them.[26] When obtaining calories from the land efficiently was an object and when units of labor expended to obtain these calories were taken into account, *Castanea sativa* amply recompensed its cultivators for the relatively little labor they invested on the tree. In ninth-century Campania, where the chronicler Erchempert claimed labor was so scarce that at the end of summer urban folk abandoned

---

[24] Fenaroli, *Il castagno*, p. 30 claimed Campanian trees produced a measly average of 7 kilos of nuts per year in the 1930s. Cherubini, "La 'civiltà,'" p. 272 gives the much higher production statistics, in tune with Pitte, *Terres de castanide*, pp. 143 and 199–200. Chestnuts' caloric contribution is discussed by Bruneton–Governatori, "Alimentation," p. 1161.

[25] Catherine Bourgeois, *Le châtaignier, un arbre, un bois* (Paris, 1992), pp. 129–31 analyzes the long term effects of chestnut coppicing, which would only deplete the soil if regular removal of leafy biomass were pursued over centuries and cutting cycles were very brief. See also Fenaroli, *Il castagno*, pp. 98–9.

[26] Pitte, *Terres de castanide*, pp. 197–201. A modern, well-tended hectare of chestnuts requires 20 days of a person's labor every year. It generates on average 1000 kilos of nuts, or a little less than 2 million edible calories.

their cities for the vineyards in order to make sure the vintage was brought in, this would have been a very attractive characteristic.[27]

Thus the European chestnut's vitality, high productivity, and long life span combined with the resurgence of woodland, the low demographic pressure of the early medieval period and the dearth of labor and lessened capacity of the powerful to coerce it, as well as with the new cultural predisposition to consider trees as economic assets, to create the Campanian landscape legible in the early medieval charters. No longer a marginal presence or an exotic plant, along the Amalfi coast and in the Cava valley that led north from the coast, *Castanea sativa* by 1000 had established itself as an unlikely dominant species.

The charter that the priest, Constantine, wrote "with his own hand" in January 1036 offers its readers a particularly dazzling glimpse into the world of southern Campania's chestnut cultivators.[28] Constantine was a well-connected individual and had several members of Amalfi's aristocratic families witness his agreement with Leo, son of Sergius of Palumola. That itself is a revealing circumstance that belies the image of chestnuts as the "bread" tree of poor and marginalized populations; it is a sign that instead chestnut trees and their fruits were not a marginal component of the local economies, but interested privileged Campanians who had other options.[29] The priest drew up the contract in his capacity as rector of the Amalfitan church of St. Mary; the property he assigned to Leo in a long-term lease, applicable to his own and to Leo's heirs "from generation to generation for all eternity" belonged to St. Mary through a donation by "lord Lupinus son of Count Maurus."[30] The land was agriculturally varied, and everyone expected it to produce diverse fruits, including figs, apples, pears, grapes, and "light cherries," in quantities large enough to warrant some charitable distribution of excess.

Two crops merited somewhat more consideration than the others, however. As this was an "ad pastinandum" contract, Leo was expected

[27] *Historia Langobardorum Beneventanorum*, ed. Georg Waitz, *MGH Scriptores Rerum Langobardicarum* (Hannover, 1878), p. 257. Admittedly the area had endured exceptional military depredations in the mid-800s.

[28] *Il codice Perris* 35, pp. 48–50. The older version destroyed in 1943 was less complete: *Codice Diplomatico Amalfitano* 44, pp. 67–8. For his edition Filangieri di Candida, who knew of the Codex Perris, preferred to use parchment originals that derived from SS. Trinità's archive and had entered the Neapolitan Archivio di Stato in the nineteenth and early twentieth centuries: Allocati, "Il cartulario," p. 364.

[29] As happened in the 1100s: Toubert, "Paysages ruraux," pp. 210–11.

[30] On the Comitis Mauronis families, see Adolf Hofmeister, "Der Übersetzer Johannes und das Geschlecht Comitis Mauronis in Amalfi," *Historische Vierteljahrschrift* 27 (1932), 256–7 and 493–4.

to improve the farm, and the wording of the deed suggests that improving the vineyard was a high priority: "[F]rom this day on you and your heirs shall take care to properly work the vineyard and stake it and plant vines in it and extend it and fence it in, as it deserves."[31] Most grapes would become wine, and Leo agreed to make the wine, aided by someone sent by St Mary. Together they would seal the wine in the church's containers (though the verb "inbuctare" implies barrels, the wine would actually rest in the thoroughly cleaned and maintained vats called "organea" kept on the farm and belonging to St Mary). Constantine required Leo and his helper to deliver a portion of this wine to the sea's shore, so we may deduce that autarchic consumption was not the destiny of the entire vintage.

The other crop to earn special attention from the contracting parties was the chestnut crop. The wood that this grove at "Insubrizzano" might yield attracted no interest, despite its obvious utility in the vineyard.[32] St Mary's chestnuts had value only for their nuts. This certainly was related to the nature of the trees, for they had been grafted at some earlier time, perhaps by Leo's father who had tended the land before, but probably much earlier. Constantine identified the chestnut trees as "inserte" and "zenzale" (as we shall see, a cultivar) and the whole grove was called an "insertetum": early medieval Campanians took very seriously the distinction between simple "castanieta" and the grafted, improved trees (or "inserte;" clustered, they formed an "insertetum"). Trees that were ungrafted, or wild, could and did produce nuts, but not as abundantly as grafted trees could, so they were better suited to producing wood. Trees that developed from a young plant onto which a shoot from a "proved" producer of bigger, or sweeter, or longer-lasting, or more abundant chestnuts had been grafted were altogether a superior agricultural prospect.[33] St. Mary's grafted chestnut grove was thus

---

[31] On contract types, Augusto Lizier, *L'economia rurale dell'età prenormanna nell' Italia meridionale* (Palermo, 1907), pp. 75–104.

[32] This locality probably lay up the Reginna torrent from Maiori, a village east of Amalfi. The *Codice Diplomatico Amalfitano* version read "in Sulfizzano," which is how other charters in the Perris Codex refer to a site in the Tramonti highlands above Maiori: *Il codice Perris* 119, p. 209 (AD 1128), and 130, p. 232 (AD 1138), the latter being the sale whereby S. Lorenzo acquired the land, explaining why the nuns kept Leo's contract.

[33] In the nineteenth and twentieth centuries prized qualities that earned chestnuts the chance to become a recognized cultivar were flavor (sweetness), consistency (silkiness), size, time of ripening, and preservability: Giovanni Vitolo, "I prodotti della terra," in *Terra e uomini nel Mezzogiorno normanno-svevo*, ed. Giosuè Musca (Bari, 1987), pp. 175–6.

an economic asset equal to the church's commercial vineyard; this begins to explain the relatively high status of some of the people involved in the contract. In fact, the same clause that had specified how Leo should improve the vines continued to state "concerning the other asset, that is the grafted chestnut grove, you must take care to prune it well [...] every year and rake and, where appropriate, to graft it and fill it in from one boundary to the other, so that there are no empty patches in it, that it may be totally full of vines, of grafted trees, of zenzale type trees, and fruit trees..."[34]

Like the grapes, the chestnuts too had to be processed. To begin, Leo and his heirs would have to gather them, probably from the ground as was customary in the Middle Ages. To facilitate this operation Constantine stipulated that the ground under the trees should be raked, and thus the undergrowth in the grove be kept low. The thick shade created by the leafy canopy of chestnut trees is inimical to many plant species but in Mediterranean climates, as we have seen, *Castanea sativa* often grows its leaves late. This meant more precocious plants could get a head start so that, by the time the chestnuts matured and fell off the trees in late autumn (around Salerno today, in November and early December, but in the early Middle Ages before the feast of St. Martin on November 3rd), a mantle of grasses and shrubs under the chestnuts might render the gathering of the crop difficult. In order for people to get a larger share of the fallen chestnuts, the prevident priest Constantine asked that Leo labor on the wild grasses beneath the trees, improving visibility there and thereby also improving yields.

Once he had gathered them, Leo had to sort the chestnuts. This Amalfitan charter is the earliest Campanian charter to identify several different chestnut types and to reveal the discernment that eaters of the nuts brought to the act. Quite apart from the distinction between "zenzale" and other "inserte," both grafted chestnuts, Constantine also knew about green grafted chestnuts, and about vallania and verole varieties.[35]

[34] *Il codice Perris*, 35, p. 49, "... de ipso alio qui est insertetum habeatis curam ad bene illud [lacuna] roccandum omni annue et rastillandum et ubi meruerit insertandum et implendum de fine in finem, ita ut vacuum ibidem non habeat set totum plenum siat de vinea et insertetum et zenzaletum et fructura..."

[35] Jean-Marie Martin, "Città e campagna: economia e società (sec. VII–XIII)," in *Storia del Mezzogiorno* 3, ed. Giuseppe Galasso et al. (Naples, 1990), p. 326 notes that "zenzala" was a cultivar Salernitans knew by 1010. See also Lizier, *L'economia*, p. 123. The cultivar "vallania" was still known to late medieval writers: Charles DuCange, *Glossarium mediae et infimae latinitatis* 6 (Paris, 1846), p. 730.

He wanted St Mary's share to include a *modium*, or about 51 kilograms, of these three types of chestnuts mixed together, and delivered directly to the church.[36] Evidently this payment was of fresh chestnuts, still with their shell on, which is why the contract specified that the delivery should occur "in chestnut season" (and, as we shall see, Constantine was quite precise about the other, dried chestnuts Leo would pay each year). In the vicinity of Avellino, twelfth-century landlords also occasionally insisted on obtaining "green" chestnuts.[37] Though it is not clear exactly what a green chestnut was in 1030s Campania, in late medieval Emilia Piero de' Crescenzi thought green chestnuts were those that one gathered before they fell off the tree, a practice that recommended itself where pigs roamed and might deprive people of their crop. Perhaps making a virtue of necessity, de' Crescenzi proposed that the green nuts were the most flavorful kind, and the kind that kept with the least risk of spoilage.[38] Indeed, superior preservation qualities appear to have been a significant benefit of immature (presumably green) chestnuts, justifying the added labor gathering them required. In the late Middle Ages, green chestnuts were stored with their spiny outer shell on, which made it possible later to eat them whole, not ground into meal.[39]

Regardless, Constantine's green chestnuts would reach him along with the other two varieties, presumably chestnut brown and ripe. Unlike the green chestnuts, which may refer to the degree of maturation they had reached when they were knocked off the branches, the other two names for chestnuts refer to cultivars, as does "zenzale." Pliny the Elder had immortalized a Roman landowner and his freedman who had likewise created new cultivars of chestnut in Campania a millennium earlier, probably as a hobby, but such scientific arboriculture is not usually associated with early medieval peasants.[40] Indeed, the existence of these cultivars in early medieval Amalfi is somewhat surprising, given that in early medieval Sabina and in Lombardy chestnut

---

[36] Montanari, *L'alimentazione contadina*, pp. 167–9 sensibly reviews these measures.

[37] Martin, "Città e campagna," p. 326. Martin was unsure why "green" chestnuts were sought after. See also Vitolo, "Il castagno," p. 32.

[38] Piero De Crescenzi, *Ruralia commoda* 5.6.5–6, ed. Will Richter, 4 vols. (Heidelberg, 1996), vol. 2, pp. 106–107.

[39] Massimo Montanari, "Un frutto ricco di storia," in *La castagna sulle tavole d'Europa* (San Piero al Bagno, 2001), pp. 60–1.

[40] *Nat. Hist.* 15.94 records the Corelliana and Tereiana cultivars, named after their inventors (see 17.122). Pliny thought one more resistant, the other finer. Pliny also (15.93–4) listed the "popularis nigra," the "coctiva," the "Salariana," the "Tarentina," and the "triangula" varieties.

cultivation appears to have depended on the encouragement of wild trees and the removal of chestnuts' competitors in the woodland.[41] Constantine's charter reveals a much more refined, arboriculturally informed management of the woods to obtain the desired quality and quantity of chestnuts. This was not just a matter of planting, transplanting, and grafting, for the mixture of several types of chestnut in the grove of St Mary at "Insubrizzano" was agronomically astute, as chestnuts from different cultivars cross-pollinate more effectively and consequently produce more nuts.[42] Likewise, the intermixture of trees that had been planted recently with the older and nut-producing chestnuts, virtually required by the terms of the contract, also contributed to the health and long-term productivity of the grove.[43]

Amalfi's chestnuts were different, then, a result of intelligent grafting procedures by Leo's predecessors working the farm. Their selection of desirable qualities, and subsequent transposition of shoots from the selected trees onto other (wild?) chestnut matrixes, by 1036 had created valuable differences in the chestnut crop, as well as in the grove itself. Such differences resulted in a vernacular knowledge and classification system that was evidently widespread enough at the beginning of the eleventh century to find its way into contracts without further explanation.[44] Everyone from the contracting parties to the witnesses understood what distinguished a verola chestnut from a zenzala or a vallania, and why a landowner would want all three in his grove and his pantry. That is a sign of the antiquity of these cultivars and suggests that the "scientific" arboricultural practices the names reflect were far older than the contract itself. They were probably not Roman, both because of the scanty evidence of Roman interest in chestnut cultivation in Campania, and because classical Latin lacked the botanical refinement displayed by the Amalfitan notaries: in Roman times people got confused because the same word covered chestnuts and walnuts.[45] The arboricultural practices revealed by the naming of cultivars were

[41] Montanari, *L'alimentazione*, pp. 37–43 and 297–9; and Toubert, *Les structures*, pp. 177–9, 190–2 and 345–7.

[42] Pitte, *Terres de castanide*, p. 171.

[43] Planting, grafting, transplanting, and pruning can kill a young tree even when climate is benign; the open spaces in the grove to which Constantine referred were likely the natural outcome of attrition.

[44] *Codice Diplomatico Amalfitano* 45, p. 70 (1036) also refers to "zinzale." This charter replicates some of the terms of Constantine's.

[45] Meiggs, *Trees and Timber*, p. 421.

instead rooted in the immediate postclassical centuries, when the Roman landscape made way for the boskier early medieval one.[46] Indeed, the genealogy of St. Mary's chestnut grove was long enough that some of it was recorded in the charter itself, through the reference to the previous owner of the trees, "lord Lupinus," and to its previous cultivator, Sergius "da Palumola," father of Leo.[47]

In fact, Constantine's charter is exceptionally valuable as a shaft of light into early medieval arboricultural practice and into the sophisticated world of postclassical Campanian chestnut cultivation. By the end of the first millennium Campanian farmers had fashioned a new kind of woodland landscape through their labor. This new woodland was not exclusively a chestnut forest, but it drew on farmers' meticulous observation of chestnut trees' characteristics, their determination of which characteristics were valuable, and their propagation of desirable characteristics by excising, grafting and planting. When Constantine began writing, a managed woodland landscape was well established on the steep slopes of the Amalfitan peninsula, one that included several cultivars of chestnut recognizable and endowed with specific names. People who inhabited this landscape in the early eleventh century applied the names to the trees in a way that implies their ancient and commonly accepted presence, the result of habits and knowledge that pre-date the charter itself.[48] This sort of arboriculture, like the creation of the plump Lombard "marroni" chestnuts, whose inner cuticle does not divide the white flesh (as is the case with normal chestnuts), is usually associated with the high Middle Ages.[49]

As surprising as the evidence in it of precocious variety selection and arboriculture is the reference in Constantine's charter to the dried chestnuts Leo owed St. Mary. Though there is no evidence that chestnuts were

---

[46] Vitolo, "I prodotti," p. 175 discusses some other cultivars named in Cavese charters. Like Martin (e.g. "Le travail agricole," in *Terra e uomini nel Mezzogiorno*, pp. 121–30) and other specialists, Vitolo considers the distinctive Amalfitan and Campanian woodland a post-1000 development.

[47] Other eleventh-century charters refer to previous ownership of groves, likewise pushing their origin back in time: *Codice Diplomatico Amalfitano* 50, p. 78; and 69, p. 110. Genealogical memory in Amalfi extended back about four generations, so many farms mentioned in tenth-century charters can be retraced to the 840s and exceptionally to the mid-700s: del Treppo, "Una città," pp. 102–5.

[48] And post-date classical antiquity that did not know these names: Jacques André, *Lexique des termes de botanique en Latin* (Paris, 1956), p. 76.

[49] Thirteenth-century Parisians prized Lombard "marroni": Roger Grand and Raymond Delatouche, *L'agriculture au moyen âge* (Paris, 1950). On their development, Montanari, *L'alimentazione*, p. 296.

preserved in Roman times, dried chestnuts (*castanee bene sicce*) are quite common in early medieval Campanian charters, and indeed in later, better documented times drying was the standard procedure among chestnut cultivators, the best means to ensure preservation of the crop. What surprises therefore is not the evocation of this product, but the prominence given to its "transport down to the sea shore." Oddly, the clause obliging Leo to carry St. Mary's share of the "properly dried" chestnuts to the sea appears to have occurred to the contracting parties after all other aspects of the agreement had already been settled. It is the very last stipulation before the witness list, and it comes after the standard concluding references to the landlord's right of eviction in the event of unsatisfactory work by the tenant, and after the establishment of penalties for any attempt to alter the terms of the contract.[50] If not actually an afterthought, the phrase requiring Leo and his heirs to deposit the church's dried chestnuts where they could be loaded onto a ship seems to reflect some last minute negotiations. The clause may have been decisive for Leo's widow Marenda who chose in 1112, seventy-six years after her husband had contracted to cultivate the chestnut grove, to relinquish the land because "we have become poor and cannot oversee and work this land and the chestnut grove." For an old woman, evidently alone, the task of bringing the chestnuts down to the sea was too much.[51]

Yet the transportation clause also certainly reflects the importance of dried chestnuts in postclassical Campanian commerce. Unlike the fresh chestnuts Constantine may have planned to eat himself, the dried chestnuts were destined for market, as their handling by Leo indicates. Drying chestnuts is laborious, far more than growing them. It requires firewood, which may be where the wood from the chestnut grove that Leo was supposed to prune annually would end up, even if that would not have sufficed to keep a smoky fire burning for two or three weeks, as was the standard practice in early modern times.[52] Keeping the fire

[50] Landlords commonly worried about getting their rents to the shore: Romualdo Trifone, "Le prestazioni degli antichi coltivatori amalfitani," *Rivista di diritto agrario* 8 (1929), 545.

[51] *Il codice Perris* 100, p. 70, "...venimus ad paupertatem et non potuimus continere et laborare ipsa suprascripta hereditatem et ipso predictum castanietum." Leo must have married Marenda late in his life for her to be alive in 1112. It is possible that Leo had contracted to farm another chestnut grove at Subrizzano, but given the nuns of S. Lorenzo's preservation of his 1036 agreement, this charter probably involves the same land mentioned in the 1036 charter, even if the owners in 1112 were not the same ones as in 1036.

[52] Martin, "Città e campagna," p. 326; and Fenaroli, *Il castagno*, pp. 113–4.

at the requisite low smoldering level was tiresome, as was removing chestnut wood's abundant ash afterwards. In more recent times such drying has taken place in specialized buildings, subdivided horizontally with racks on which the chestnuts lie above a slowly burning fire. Constantine's charter actually records one of these structures, surely one of the earliest known in the Italian peninsula.[53] In his description of the farm he leased to Leo, Constantine reserved for St. Mary's exclusive use "those barrels and that house which I recently built there for the racks of the aforementioned chestnut grove..." The rack-house was again mentioned as the site for dividing the crop into shares, under the vigilant eye of someone sent by St. Mary, suggesting that it was a special structure on the farm, where the landlord's power over the operation of farm work could be manifested. The division took place after drying; this implies that what was divided were fully peeled nuts, for drying the nuts made their separation from outer shell and inner cuticle a simple operation, one done immediately after drying in early modern times. That is what Constantine meant by "properly dried" chestnuts, those that Leo would carry down to the Tyrrhenian, where the landlord would make sure to have them met by boats. The now preservable, lighter, less voluminous, and hence commercial chestnuts could be distributed to the right place by sea.[54]

The agreement between Leo and Constantine is therefore another sign of coastal Campania's commercial vitality at the end of the first millennium. This corner of the Tyrrhenian was perhaps the liveliest part of the early medieval Mediterranean, where landlords shipped surpluses to markets as a matter of course, from the ninth century (when the documentation thickens) on.[55] Even as early as 836, government documents like the *Sicardi Pactio cum Neapolitanis* evince an unusual concern with the movement of goods and people across boundaries in the

---

[53] Vitolo, "Il castagno," p. 30 discusses some later racks.

[54] Drying removes a third of chestnuts' weight: Fenaroli, *Il castagno*, p. 105. Vitolo, "I prodotti," p. 177 links chestnut cultivation to Maghrebi market demand. Salerno and Naples are likelier markets. P. Bevilacqua, *Terre del grano, terre degli alberi: L'ambiente nella storia del Mezzogiorno* (Rionero, 1992), pp. 77–80 remarked on southern arboriculture's "dynamism" around 1900, when coastal production for export predominated, in contrast with inland "static" grain growing.

[55] Chris Wickham, *Framing the Early Middle Ages* (Oxford, 2005), pp. 737–41. See also P. Skinner, "The Tyrrhenian Coastal Cities Under the Normans," in *The Society of Norman Italy*, ed. G. Loud and A. Metcalfe (Leiden, 2002), 90–92; and G. Loud, "The Monastic Economy in the Principality of Salerno During the Eleventh and Twelfth Centuries," *Papers of the British School at Rome* 71 (2003), 169–70.

region.[56] This was also a part of the Italian peninsula where that shy tree *Castanea sativa* had become a major economic protagonist. At about the same time as Leo and Constantine were negotiating their terms, viticulture in Campania completed its steady rise to prominence as the preferred, most commercially viable investment for landowners. Southern Campanian vineyards did not, however, extirpate or marginalize chestnuts. If chestnut groves remained a viable presence in the wooded landscape, even at low altitudes, into the 1200s, this was probably not solely because they supplied viticulturalists with high quality poles. Though landlords began to replace their chestnut woods in the 1200s, the trees' resiliency and continued importance in the high medieval landscape derives from some of the patterns the 1036 charter suggests, namely the high degree of sophistication chestnut cultivation had attained in this region, the involvement in such cultivation of powerful people, the Campanian taste for high quality nuts, and the ability of dried chestnuts to serve as a cash crop. The managed woodland of the Amalfi peninsula, like that in Campania generally, was a sustainable response to market conditions. It was also a result of the remarkable properties of *Castanea sativa*, an unassuming tree that gives generations of people abundant fruit in exchange for minimal maintenance of its ecosystem.

## *Bibliography*

Allocati, Antonio. "Il cartulario amalfitano detto comunemente 'Codice Perris,' e la sua edizione," in *Convegno internazionale 14–16 giugno 1973. Amalfi nel medioevo* (Salerno, 1977), pp. 361–65.

André, Jacques. *Lexique des termes de botanique en Latin* (Paris, 1956).

Barbiera, Irene and Gianpiero Dalla Zuanna, "Le dinamiche della popolazione nell'Italia medievale," *Archeologia medievale* 34 (2007), 19–42.

Becchi, Massimo. *Discorso sul castagno* (Reggio Emilia, 1996).

Beloch, Karl. *Bevölkerungsgeschichte Italiens*, 3 vols. (Berlin, 1937–61).

Bevilacqua, Piero. *Terre del grano, terre degli alberi. L'ambiente nella storia del Mezzogiorno* (Rionero, 1992).

Bourgeois, Catherine. *Le châtaignier, un arbre, un bois* (Paris, 1992).

Bruneton-Governatori, Ariane. "Alimentation et idéologie," *Annales: Economies, société, civilisations* 39 (1984), 1161–89.

Conedera, Marco, et al. "Competition and Dynamics in Abandoned Chestnut Orchards in Southern Switzerland," *Ecologia Mediterranea* 26 (2000), 101–12.

Cherubini, Giorgio. "La 'civiltà' del castagno in Italia alla fine del medioevo," *Archeologia medievale* 8 (1981), 247–80.

[56] *Sicardi Pactio cum Neapolitanis*, in *MGH Legum* 4, ed. Georg Heinrich Pertz (Hannover, 1868), pp. 217 and 219–20.

Christie, Neil. *From Constantine to Charlemagne: An Archaeology of Italy, AD 300–800* (Aldershot, 2006).
De Crescenzi, Piero. *Ruralia commoda*, ed. Will Richter, 4 vols. (Heidelberg, 1996).
Del Treppo, Mario. "Una città del Mezzogiorno nei secoli IX–XIV," in *Convegno internazionale 14–16 giugno 1973. Amalfi nel medioevo* (Salerno, 1977), pp. 19–175.
Devroey, Jean-Pierre. *Économie rurale et société dans l'Europe franque*, 2 vols. (Paris, 2003–05).
Fenaroli, Luigi. *Il castagno* (Rome, 1945).
Filangieri di Candida, Riccardo, ed. *Codice Diplomatico Amalfitano* (Naples, 1917).
Giovannini, Fabio. *Natalità, mortalià e demografia dell'Italia medievale* (Oxford, 2001).
Grand, Roger and Raymond Delatouche. *L'agriculture au moyen âge* (Paris, 1950).
Grove, A. T. and Oliver Rackham. *The Nature of Mediterranean Europe* (New Haven, 2001).
Hoffmann, Richard C. "Medieval Christendom in God's Creation," in *Northern Europe: An Environmental History*, ed. Tamara Whited et al. (Santa Barbara, Denver and Oxford, 2005), pp. 45–72.
Hofmeister, Adolf. "Der Übersetzer Johannes und das Geschlecht Comitis Mauronis in Amalfi," *Historische Vierteljahrschrift* 27 (1932), 225–84, 493–508, 831–3.
Horden, Peregrine and Nicholas Purcell. *The Corrupting Sea: A Study of Mediterranean History* (Oxford, 2000).
Jashemski, Wilhelmina et al. "Catalogue of Plants," in *The Natural History of Pompeii*, ed. Wilhelmina Jashemski and Frederick Meyer (Cambridge, 2002), pp. 94–180.
Lizier, Augusto. *L'economia rurale dell'età prenormanna nell' Italia meridionale* (Palermo, 1907).
LoCascio, Elio and Paolo Malanima. "Cycles and Stability: Italian Population Before the Demographic Transition," *Rivista di storia economica* 21 (2005), 197–232.
Loud, G. "The Monastic Economy in the Principality of Salerno During the Eleventh and Twelfth Centuries," *Papers of the British School at Rome* 71 (2003), 141–79.
Lowe, J. John, et al., "Stratigrafia pollinica olocenica e storia delle risorse boschive dell'Appennino settentrionale," *Rivista geografica italiana* 102 (1995), 267–310.
Martin, Jean-Marie. "Le travail agricole," in *Terra e uomini nel Mezzogiorno normanno-svevo*, ed. Giosuè Musca (Bari, 1987), pp. 113–57.
——. "Città e campagna: economia e società (sec. VII–XIII)," in *Storia del Mezzogiorno*, ed. Giuseppe Galasso et al., 15 vols. (Naples, 1990), vol. 3, pp. 259–382.
——. "Settlement and the Agrarian Economy," *The Society of Norman Italy*, ed. G. A. Loud and Alex Metcalfe (Leiden, 2002), pp. 17–45.
Mazzoleni, Jole and Renata Orefice, ed. *Il codice Perris* (Amalfi, 1985).
Meiggs, Russell. *Trees and Timber in the Ancient World* (Oxford, 1982).
Mols, Stephanus. *Wooden Furniture in Herculaneum* (Amsterdam, 1999).
Montanari, Massimo. *L'alimentazione contadina nell'alto medioevo* (Naples, 1979).
——. *Campagne medievali* (Turin, 1984).
——. "Dalla tardoantichità all'alto medioevo," in *Storia dell'alimentazione*, ed. Jean-Louis Flandrin and Massimo Montanari (Bari, 1997), pp. 211–33.
——. "Un frutto ricco di storia," in *La castagna sulle tavole d'Europa* (San Piero al Bagno, 2001), pp. 50–63.
Pinto, Giuliano and Eugenio Sonnino. "L'Italie," in *Histoire des populations de l'Europe*, ed. Jean-Pierre Bardet and Jacques Dupâcquier (Paris, 1997), pp. 485–508.
Pitte, Jean-Robert. *Terres de castanide* (Paris, 1986).
Pugsley, Paola. *Roman Domestic Wood* (Oxford, 2003).
Quirós Castillo, Juan Antonio. "Cambios y transformaciones en el paysaje del Appennino toscano entre la Antigüedad Tardía y la Edad Media," *Archeologia medievale* 25 (1998), 177–97.
Renn, Jürgen and Giuseppe Castagnetti, ed. *Homo Faber* (Rome, 2003).
Rouche, Michel. "Le haut moyen âge," in *Histoire des populations de l'Europe*, ed. Jean-Pierre Bardet and Jacques Dupâcquier (Paris, 1997), pp. 133–67.

Skinner, Patricia. "The Tyrrhenian Coastal Cities Under the Normans," in *The Society of Norman Italy*, ed. G. A. Loud and Alex Metcalfe (Leiden, 2002), pp. 75–96.
Tropeano, Placido, ed. *Codice Diplomatico Verginiano*, 13 vols. (Montevergine, 1977–2000).
Toubert, Pierre. *Les structures du Latium médiévale* (Rome, 1973).
——. "Paysages ruraux et techniques de production en Italie méridionale dans la seconde moitié du XIIe siècle," in *Potere, società e popolo nell'età dei due Guglielmi* (Bari, 1981), pp. 201–29.
Trifone, Romualdo. "Le prestazioni degli antichi coltivatori amalfitani," *Rivista di diritto agrario* 8 (1929), 541–51.
——. *Storia del diritto forestale in Italia* (Florence, 1957).
Vitolo, Giovanni. "I prodotti della terra." in *Terra e uomini nel Mezzogiorno normanno-svevo*, ed. Giosuè Musca (Bari, 1987), pp. 159–85.
——. "Il castagno nell'economia della Campania medievale," *Rassegna storica salernitana* 12 (1989), 21–34.
Wickham, Chris. "European Forests in the Early Middle Ages," in *Settimane di studio del centro italiano di studi sull'alto medioevo* 37 (Spoleto, 1990), pp. 479–545.
——. "Agricultura, ambiente e sviluppo economico nella storia europea: il problema dell'alto medioevo," in *Agricultura ambiente e sviluppo nella storia europea*, ed. Luciano Segre (Milan, 1993), pp. 157–61.
——. *Framing the Early Middle Ages* (Oxford, 2005).
Williamson, Tom. *Shaping Medieval Landscapes* (London, 2003).

# THE GREAT FAMINE: 1315–1322 REVISITED*

William Chester Jordan

In 1996 I published a book on the Great Famine of the early fourteenth century, a phenomenon that encompassed all of northern Europe. Since the publication of the book, or, rather, from the time I finished writing it nearly two years before, several new sources and studies have appeared in print. This paper employs a selection of these materials that I have not treated elsewhere in order to further our knowledge and understanding of the causes, experience and consequences of the Great Famine, which, excepting only the Black Death of 1347–52 and the plague cycle it initiated, was the greatest natural catastrophe of the Middle Ages.[1]

* * *

The subject of famine has generated an enormous bibliography – historical, biological, social, political, literary, and economic. The sources from which scholars have constructed their narratives and commentaries vary from period to period, and the governing paradigms shift according to the same criterion. The sources on the Great Famine of the early fourteenth century are extensive and varied. The basic narrative line is drawn from a very large number of chronicles (*GF*, 7–10). The event lasted so long and was so severe that nearly all contemporary chroniclers (monks, generally speaking) mentioned it and devoted a great deal of space to it. Much of these writers' commentary

---

* I presented a version of this paper in Florence on 8 March 2007 to the European University Institute Seminar on Economic History (directed by and at the invitation of Professors Harold James and Giovanni Federico). I wish to thank the organizers and the participants most warmly.

[1] In the text I give abbreviated references to the fuller discussions in the book (*GF* for *The Great Famine: Northern Europe in the Early Fourteenth Century* [Princeton: Princeton University Press, 1996]) and in a subsequent article (F&P for "Famine and Popular Resistance: Northern Europe, 1315–1322," in *Power, Violence and Mass Death in Pre-Modern and Modern Times*, ed. Joseph Canning, Hartmut Lehmann and Jay Winter [Aldershot, UK, 2004], pp. 13–24). In the following notes, I ordinarily make reference to materials not consulted for these earlier publications.

as opposed to their reporting of 'facts' employs tropes from the bible's descriptions of ancient famines (*GF*, 148–49), but it is possible to disarticulate the two forms of information. The biblical tropes and the precise forms of their use are sources in their own right of attitudes towards the catastrophe (*GF*, 149–50). A recently published fourteenth-century chronicle of particular value is that conventionally attributed to John Somer, an English Franciscan. In fact, Brother John owned one of the manuscripts of the chronicle but may not be the author. Whether the matter of authorship will ever be resolved or not, the chronicle, which for convenience I will be calling Somer's *Chronicle*, provides abundant information verifiable sometimes from other sources, which increases historians' confidence in its various descriptions.[2]

Besides chronicles there are decrees of royal and municipal governments regulating access to resources, responding to rising prices, and threatening criminals, moneylenders, and hoarders who exploited conditions (*GF*, 158–59, 168–69, 171–73). Synodal decrees express bishops' and priests' concerns with poverty and alms giving (*GF*, 172). Letters record the observations and particular concerns of a selection of government and ecclesiastical officials (*GF*, 76, 172). Data on wages and prices and on production are preserved in thousands of parchment manuscript pages of estate accounts, mostly monastic (*GF*, chaps. 4–6). Among those recently published is a selection of documents of Westminster Abbey's Hertfordshire landholdings, which turn out to have much to say about the interplay of the weather and agricultural management, and the Bolton Priory *computus* rolls, which Ian Kershaw partly exploited from the manuscripts in his now thirty-year old but still wonderful and extraordinary study of northern English Famine conditions.[3] Unique to England, manorial court records, like those of Downham manor in East Anglia, provide evidence of disputes that give an insight into social and economic relations in the period.[4] Charters inscribe

[2] "The Chronicle of John Somer, OFM," ed. Jeremy Catto and Linne Mooney, in *Chronology, Conquest and Conflict in Medieval England*, Camden Miscellany, 5th series, 10 (Cambridge, 1997), pp. 197–285.

[3] Derek Stern, *A Hertfordshire Demesne of Westminster Abbey: Profits, Productivity and Weather*, ed. Christopher Thornton (Hatfield, UK, 2000); see especially pp. xlvii–xlviii and 94, 98–99. *The Bolton Priory Computus, 1286–1325* was edited by Ian Kershaw and David Smith (Woodbridge, 2000). Kershaw's study is "The Great Famine and Agrarian Crisis in England (1315–1322)," *Past and Present* 59 (1973), 3–50.

[4] *Court Roll of the Manor of Downham 1310–1327*, ed. M. Clare Coleman (Cambridge, 1996). On the uniqueness of English manorial records, see Zvi Razi and Richard Smith, "The Origins of the English Manorial Court Rolls as a Written Record: A Puzzle,"

brokered agreements that can also provide insight into the social and tenurial impact of the Famine; the cartulary or charter-collection book of Pakenham, recently published, is a nice example.[5] Rent rolls and annuity lists, chiefly municipal, and reports on vagabondage provide information on markets, like housing and labor, also affected by the conditions of the Famine (*GF*, 153). Requests for reimbursement for seeing to the burial of famine mortalities give some indication of the death rate (*GF*, 146–47). Archeological material provides additional and quite varied information (*GF* 37, 98). Tree-ring evidence (dendrochronology) provides a check against and indeed confirmation of chroniclers' reports of heavy rain and drought, since oaks are especially growth-sensitive to these conditions, as manifested in the width of seasonal ring-growth (*GF*, 17). Liturgical, devotional, and amuletic texts provide evidence of the religious ceremonies and quasi-religious practices available to people to sustain their emotional health and hope (*GF*, 108–09, 156–84).

I have already remarked that the Famine affected all of northern Europe – in modern terms the British Isles, Scandinavia, northern France, the Low Countries, Germany, Austria, and the borderlands of Bohemia and Poland (*GF*, 7–8).[6] The total population of these regions on the eve of the Famine was about thirty million, roughly 90% of whom, depending on how one defines a town (F&P, 14), lived in the countryside (*GF*, 8).[7] The urban population was unevenly distributed. Norway's urban growth had leveled off by 1200, unlike on the continent where the population of towns and the number of towns continued their increase down to 1300. No more than about 20,000 of Norway's 400,000 people (5%) could be considered urban dwellers.[8]

---

in *Medieval Society and the Manor Court*, ed. Zvi Razi and Richard Smith (New York, 1996), pp. 36–68.

[5] *Pakenham Cartulary for the Manor of Ixworth Thorpe, Suffolk, c. 1250–c. 1320*, ed. S. D. Church (Woodbridge, 2001).

[6] See also Nils Hybel, "Klima og hungersnød i middelalderen," *Historisk Tidsskrift* (Denmark) 102 (2002), 265–81 (with English summary).

[7] David Nicholas, *Urban Europe, 1100–1700* (New York, 2003), pp. 6–10; Volker Stamm, "Gab es eine bäuerliche Landflucht im Hochmittelalter? Land-Stadt-Bewegungen als Auflösungsfaktor der klassischen Grundherrschaft," *Historische Zeitschrift* 276 (2003), 308; Christopher Dyer, "The Urbanising of Staffordshire: The First Phases," *Staffordshire Studies* 14 (2002), 1–31; and Howard Clarke, "Decolonization and the Dynamics of Urban Decline in Ireland, 1300–1500," in *Towns in Decline AD 100–1600*, ed. T. R. Slater (Aldershot, UK, 2000), p. 163.

[8] Richard Holt, "What If the Sea Were Different? Urbanization in Medieval Norway," in *Rodney Hilton's Middle Ages*, ed. Christopher Dyer, Peter Coss and Chris Wickham (Oxford, 2007), pp. 139–41.

The British Isles had many small towns but only one large city (large by medieval standards), London, estimated at 80,000 or more according to recent speculation, more like 40,000 according to the traditional scholarly consensus (*GF*, 128).[9] Northern France had Paris, about 200,000, the largest city in Western Europe (*GF*, 131), larger even than Florence.[10] Northern France and Flanders, the most urbanized parts of northern Europe at the time, had several cities in the range of population 30,000–60,000 (*GF*, 131), including Tours, Rouen, Saint-Omer, Ghent and Bruges.[11] The Rhineland metropolis, Cologne, had 50,000 residents according to the most recent estimate (cf. *GF*, 130).[12] The North Sea and Baltic trading towns like Hamburg and Lübeck were smaller, 5,000–20,000 (*GF*, 130–31). Despite the relatively low populations of these towns, their surface area was so small that they were densely populated by modern standards, and their inhabitants were far more susceptible to epidemic disease than were their rural counterparts.

The foregoing problems notwithstanding, the thirteenth century had been good to both the rural and urban dwellers of northern Europe. Population had never been higher (or healthier, according to the evidence of skeletal remains), and its increase had been sustained by steady economic growth (*GF*, 11–12, 128–31; F&P, 14, 17).[13] Warning signs around the year 1300 of stress on resources, however, included the tapering off of economic growth, geographically localized recessions, increased pressure on common rights, inadequate credit, dangerous depletion of woodland resources, especially but not exclusively near towns, an incapacity to expand arable further into coastal wetlands, the cultivation of economically marginal lands and those of relatively low natural fertility, and, finally, the likely beginning of a decline in overall per capita consumption (*GF*, 90–95).[14] As H. M. Dunsford

---

[9] Caroline Barron, *London in the Later Middle Ages: Government and People, 1200–1500* (Oxford, 2004), p. 238; and Nicholas, *Urban Europe*, p. 13.

[10] For the comparison, see Nicholas, *Urban Europe*, p. 13. Much has been written directly on Florentine population, but see now W. R. Day, Jr., "The Population of Florence before the Black Death: Survey and Synthesis," *Journal of Medieval History* 28 (2002), 93–129, where the city population proper is estimated at 120,000; the contado had an additional 300,000.

[11] Nicholas, *Urban Europe*, p. 13; Alain Derville, *Saint-Omer des origines au début du XIVe siècle* (Lille, 1995), p. 184.

[12] Nicholas, *Urban Europe*, p. 13.

[13] On the evidence of the inhabitants' health, see Elizabeth Carpentier and Michel Le Mené, *La France du XIe au XVe siècle: Population, société, économie* (Paris, 1996), p. 59.

[14] On woodlands, Bettina Borgemeister offers a useful reminder that "Die mittelalterlichen und frühneuzeitlichen Städte lebten aus dem Wald" ("Die Bedeutung des Waldes für die Stadt," *Niedersächsisches Jahrbuch für Landesgeschichte* 78 [2006], 17).

and S. J. Harris remind us, not every danger sign appeared with equal visibility everywhere.[15] Yet, the more or less general conjuncture of factors made escape from the growing malaise difficult without an exogenous shock; hence, the use of the now widely accepted term *blocage* (*GF*, 95–96) to describe it.[16]

Pre-modern famines were largely brought on by nature-induced crises in the production of cereals (staple crops) and other food sources. They were exacerbated by other factors, like war, rebellion, piracy, the erection of new tariff barriers, and the interruptions in trade and migration that followed from these, but, unlike the case in modern times, none of these factors alone, except the long-term siege of a town, could ordinarily bring on a famine in the absence of a nature-induced crisis in production (*GF*, 12–15). Nor, obviously, was the totalitarian goal either of autarky, understood as isolation from democratic political currents and insulation from liberal (economic) ideology, or of breakneck speed industrial development relevant as a causative factor in the Middle Ages, as it has proved to be in recent times (the Soviet Union under Stalin, Communist North Korea).

The crisis in production that caused the Great Famine of 1315–22 had its proximate cause in an anomalous run of severe winters and rainy springs, summers and falls or alternatively, in the later years, equally severe winters but spring, summer and fall droughts (*GF*, 15–19; F&P, 14). "Rain," by the way, means heavy downpours day after day, according to one count, one-hundred and fifty days of such rain in a row in one year, enough to wash away seeds and seedlings, cause grain to rot, wreck fish traps, wash away dikes and levees, flood meadows, weaken tree root systems and bring down mature growth that then sometimes clogged waterways, destroy bridges, make roads impassable, and inundate villages and towns. The weather was exacerbated by other factors, war in particular and the draconian taxation that accompanied

---

See also Werner Rösener, "Der Wald als Wirtschaftsfaktor und Konfliktfeld in der Gesellschaft des Hoch- und Spätmittelalters," *Zeitschrift für Agrargeschichte und Agrarsoziologie* 55 (2007), 14–31; and J. D. Hurst, "Fuel Supply and the Medieval Salt Industry in Droitwich," *Transactions of the Worcestershire Archaeological Society*, 3rd series, 19 (2004), 111–32. On exploitation of wetlands, see Stephen Rippon, *The Transformation of Coastal Wetlands: Exploitation and Management of Marshland Landscapes in North West Europe during the Roman and Medieval Periods* (Oxford, 2000), p. 186.

[15] Cf. H. M. Dunsford and S. J. Harris, "Colonization of the Wasteland in County Durham, 1100–1400," *Economic History Review* 56 (2003), 34–56.

[16] Carpentier and Le Mené, *France du XIe au XVe siècle*, pp. 313–60.

it (*GF*, 19–21, 176–77; F&P, 15). Worse still, rulers acted as if they looked upon the Famine as God's retribution for not winning their wars rather than as a justification for ceasing to fight them. Thus, they continued to mount armies as best they could even as the Famine raged (*GF*, 20).

In northern continental Europe and in England there were two harvests a year, the so-called winter wheat harvest in the spring when plants growing from seeds sown in the fall bore grain, and the autumn harvest, when the bounty of plants grown from seeds sown in the spring was harvested. The weather made the routine tasks of husbandry – plowing, harrowing, seeding, secondary harrowing, weeding, harvesting, and gleaning of cereals; pasturing of herds and flocks – much more difficult and ineffective when they were accomplished at all (*GF*, 24; F&P, 16). The court rolls of Downham manor at a session in February 1316 make reference to farmers' inability to use wagons, wheeled cartage, for a full three weeks of the summer and fall before, presumably at the critical time of harvest or it would not have been worth remembering in court. The cause? The mud, one infers, occasioned by the terribly rainy weather (*per intemperiem temporis*).[17]

Another consequence of the weather was the fall-off in yields (*GF*, 31–33) of all cereals (*bladum*), sometimes to such a degree that producers had insufficient surplus to benefit from the high prices and thus offset the economic impact of the fall-off.[18] Vivid phrases evoke the barren return of the land in Bavaria (*propter malum satum terre ac sterilitatem*) and the pit of usury (*voraginem usurarum*) into which producers who were trying to survive plunged headlong in 1316.[19] The residual impact of this situation was volatility in the land market, though this varied significantly in intensity.[20] A second consequence of the foul weather was the spread of disease among domesticated food animals, their susceptibility having been increased both by weather-related conditions (hypothermia in winter; fluke-worm infestation during periods of excessive dampness; pregnancy toxemia during times of drought or cold wet) and by low stores and therefore low intake of

---

17 *Court Roll of the Manor of Downham*, p. 66, no. 241.

18 "Chronicle of John Somer," p. 272; and David Stone, *Decision-Making in Medieval Agriculture* (Oxford, 2005), pp. 29 and 37.

19 Alfred Wendehorst, *Das Bistum Eichstätt, I: Die Bischofsreihe bis 1535* (Berlin and New York, 2006), pp. 141–42.

20 Stone, *Decision-Making*, p. 29; and Mark Page, "The Peasant Land Market on the Bishop of Winchester's Manor of Farnham, 1263–1349," *Surrey Archaeological Collections* 90 (2003), 168.

nutritional fodder (*GF*, 35–39). *Murrain*, death by disease, is widely attested. My speculation that the principal killer was Rinderpest may or may not prove true (*GF*, 36), but what genuinely matters is not the accuracy of retrospective diagnosis but the fact of extremely heightened animal mortality. Somer's *Chronicle* tells us, for example, of epizootic diseases and widespread mortality among bulls and cows, oxen (castrated bulls), and young cattle.[21] Substitution of products from pigs, less susceptible to disease, was a short time help (*GF*, 36, 55–56), but long-term storage of meat protein was reduced by a crisis in salt production that was incidental to the inundating rains (*GF*, 52–54), a point also made in Somer's *Chronicle*. There was "such flooding in France," the author wrote under the year 1315, "that for a two-year period it destroyed (*destruxit*) salt" production.[22]

The human effects of these conditions were physical, emotional and economic. The fall-off in yields, typically in the period 1315–22 by about one-fourth to one-third, depending on the cereal, with wheat faring worse than barley, oats and rye, but none doing well (*GF*, 31–33), meant that, except for the fairly wealthy, high prices led to reduced purchases, restricted intake and nutritional disease. (Storage in the Middle Ages was insufficient to cover crop failures of this magnitude for more than two harvests [*GF*, 14–15].) Grapes and therefore wine production were also severely reduced in France and western Germany by plant diseases associated with excessive dampness (*GF*, 33–35).[23] One can add to this documented or presumed declines in the yields of peas and beans, wax and honey, fruit, and, owing to domestic animal disease, milk, butter, cheese and by-products (*GF*, 32, 54).

The impact of the nutritional exigency was varied, however, to judge both by what is deducible from contemporary sources and from the evidence of modern famines (*GF*, 115–17, 142). It was far worse for the poor in towns than for the poor on farms, who could hold back food resources for their own consumption. It was probably worse for children and the elderly than for young and middle-aged men and women. It was also probably worse for such women than for their male counterparts, for women in famine times have tended to apportion larger

---

21 "Chronicle of John Somer," p. 237: "hoc anno [1319] et sequenti fuit morina taurorum vaccarum bovum et vitulorum per totam angliam tanta quod in regno toto anglie fuit illa species animalium quasi de [leta]."

22 "Chronicle of John Somer," p. 272.

23 "Chronicle of John Somer," p. 272.

amounts of food to working husbands and brothers than to themselves. Neither in towns nor in the countryside did the wealthy endure a major food emergency, but in the towns they did suffer when pathogens occasionally jumped the barrier of class and in crowded medieval cities transmitted the diseases of the malnourished to them.

The emotional toll? The one great pillar medieval Europeans of this period had was their religion. Medieval Catholicism, which has suffered relentless criticism since the Protestant Reformation and the Enlightenment for its alleged doctrinal excesses and its obscurantism, was, we now know, a very successful religion with regard to sustaining the emotional health of the Christian people.[24] The picture of a persecuting medieval Catholicism is accurate, of course, to some degree, but we would do well to remind ourselves that the medieval Inquisition's repression of heresy was theologically major but demographically minor. Most people were good Catholics in the sense of observant of Catholic rites. Even if those rites occasionally look to us and looked to some scholastic intellectual critics of the period like old pagan or immemorial rustic ceremonies, it would be hasty to conclude that the people themselves were pagans or heathens. These old ceremonies had effectively been domesticated to the Catholic tradition. When dangers like famine came upon a believer, it was proverbial wisdom that they came as punishments from God.[25] Believers turned to the prayers and observances that promised to bring God's mercy. There were ceremonies for sheep and pigs and their feed.[26] There were rites for horses, including prayers to Saint Eligius (Eloi), the patron of metal working and therefore of the men who made horseshoes and nails.[27] The risk and occasional reality of famine had generated a large body of rites intended to assuage God's wrath and restore the natural order, a fact which may explain why non- or only half-heartedly Christian utopian

[24] See, now, for example, Robert Swanson, *Religion and Devotion in Europe, c.1215–c.1515* (Cambridge, 1995); and Augustine Thompson, *Cities of God: The Religion of the Italian Communes, 1125–1325* (University Park, Pennsylvania, 2005).

[25] "Fames: Si quando fames et penuria et omnium rerum egestas opprimit mundum, sciamus hoc ex ira dei descendere"; the sentiment in this form is reported in *Der Wolfenbüttler "Rapularius"*, ed. Hildegund Hölzel-Ruggin, MGH Quellen zur Geistesgeschichte des Mittelalters 17 (Hannover, 2002), p. 342, no. F 108.

[26] See William Chester Jordan, "Charms to Ward Off Sheep and Pig Murrain," in *Medieval Christianity in Practice*, ed. Miri Rubin (Princeton and Oxford, 2009), pp. 67–75.

[27] Ludovic Notte, "La Dévotion à saint Eloi dans les écuries princières (XIIIe–XVIe siècle," *Revue belge de philologie et d'histoire* 81 (2003), 1051–74.

fantasies, like Cockaigne, though they existed, were less prevalent than we might otherwise have suspected in the Middle Ages.[28] The possibility of a truly Catholic utopia seemed realistic even in the worst of times.

So, charitable work and alms were common, if ultimately insufficient (*GF*, 108–12, 156–62; F&P, 16, 19) – insufficient even when the effort was truly colossal, as when a Parisian furrier stipulated sufficient handouts for twenty thousand poor people at his funeral (1316), another Parisian bourgeois established a hospice to treat the sick (1318), and a third group, the squirrel curriers of Paris, founded a mutual aid society to provide steady income to families with sick breadwinners.[29] True, there were instances where personal, familial, and institutional charity broke down sufficiently during the Great Famine to undermine social solidarities and cross-class moral reciprocity (*GF*, 108–14, 162–66). Even on these occasions, such as the violent rebellion called the Crusade of the Shepherds (1320) in France and the massacres of lepers (1321) in France and its borderlands (*GF*, 170–71; F&P, 24), which are also remarked in Somer's *Chronicle*, it is not clear that the dearth so much caused the breakdowns as sharpened their intensity.[30] To put it another way and with a quite different example, somehow we have been taught to expect grain and bread riots in all pre-modern famines. I expected to find them, but, as a recent book shows, my desire to find them pushed me to overemphasize what little evidence there is on the subject.[31]

The effects of the Famine in the towns was undoubtedly greater than in the countryside, in the first instance because farmers could hold back the food crops that they needed to sustain their families. It is true that they might reduce their stores to very low levels in order to benefit from the high prices in urban food markets, and this could give rise to malnutrition without the obvious physical signs of starvation, but starvation in the technical sense (hunger disease morbidity) was characteristic of urban not rural settings. Mortality, up to 10% in some towns, was a consequence not only of hunger disease but also of opportunistic infections affecting weakened people and of epidemics (*GF*, 142–48).

---

[28] Cf. Herman Pleij, *Dreaming of Cockaigne: Medieval Fantasies of the Perfect Life*, trans. Diane Webb (New York, 2001).

[29] Sharon Farmer, *Surviving Poverty in Medieval Paris: Gender, Ideology, and the Daily Lives of the Poor* (Ithaca, 2002), pp. 33, 87 and 101.

[30] "Chronicle of John Somer," p. 272.

[31] Samuel Cohn, Jr., *Lust for Liberty: The Politics of Social Revolt in Medieval Europe, 1200–1425, Italy, France, and Flanders* (Cambridge, MA, 2006), p. 271, n. 84.

Prices climbed steeply as production in the countryside and distribution to urban markets fell, but, typical of grain markets in particular, rumors of cereals in transport to a city or predictions that the next harvest would be improved could cause prices to plummet (*GF*, 43–60, 134–37). Then an act of piracy preventing the landing of grain might send prices through the roof. Price volatility was characteristic of continental urban markets and in England (*GF*, 50, 135) in these times, but was even more severe in Scandinavia, where the levels of economic integration and inter-regional trade in foodstuffs were lower.[32] Middlemen naturally tried to control resources in such a way as to keep prices high. Their hoarding or holding back, which was resented, nonetheless had a mid-term positive effect in forcing consumers to cut back, thus allowing low stores to last over a more extended period than typical (*GF*, 49, 136–37, 168, 172). Yet, in the long run, the perdurance of the conditions that were undermining production led consumers to consume so little that they put themselves at risk of disease. To meet the challenge of high prices townspeople of the middling sort capitalized a portion of their resources – large houses for smaller houses or for lodgings, reducing inventories and leasing or selling storage space, selling back-up tools – and they reduced their number of servants in their homes and workers in their shops. The poor pawned or sold tools, utensils, dishes, linens, best clothes, and heirlooms; they cut consumption of wood; they resorted to begging for money or to putting themselves on municipal and ecclesiastical charity. Once again, resentment – from those who borrowed or pawned – of those who had the resources to lend money added to the social tension (*GF*, chap. 8).

It was not merely prices for cereals that increased steeply and endured volatility. Salt doubled in price, while wax, fruit, cider, and unfinished iron and steel showed more modest increases (*GF*, 54). Oxen, cows and sheep, goats, swine, meat by-products (grease and fat), fowl, eggs, rabbits, fish, pigs, and horses all peaked in price in the Famine (*GF*, 55–58). Some finished goods, like horseshoe nails, also saw unusual price increases (*GF*, 55). Wood which was vended in weekly urban markets

[32] A remark under the year 1318, somewhat difficult to parse, in the "Chronicle of John Somer," p. 272, may also suggest price volatility: "Post quadregisimam caristia vini et bladi fuit in (hibernia) francia per 2 annos. Autem circa pentecostem copia creavit [?] ut per vilitatem." For Scandinavia, see Bo Franzén and Johan Söderberg, "Svenska spannmålspriser under medeltiden i ett europeiskt perspektiv," *Historisk Tidskrift* (Sweden) 126 (2006), 186–214 (with English summary).

must have seen a decline both in supply, thus raising unit prices, and in purchasers, a factor that would have moderated the price rises, though probably without balancing them out.[33] Even though Europe may be said to have had a glutted labor market immediately before the Famine, wages soared in the relatively small number of geographical areas and manufacturing sectors where a labor shortage accompanied the bad weather conditions (*GF*, 59–60).

Piracy was rampant – in the Baltic Sea, the North Sea, the Channel, along the French Atlantic Coast (*GF*, 161, 168, 174–76; F&P, 17–18), and obviously must have disrupted the livelihood of towns, like Bremen, that survived in part from long-distance seaborne trade.[34] War was ubiquitous (*GF*, 19–21). In Germany there was civil war; in Norway, Denmark, and Sweden, dynastic.[35] France was at war with Flanders. England was at war with Scotland. Scotland invaded Ireland to carry the war to the English colonizers there. The Welsh were in revolt against the English. Reports of the armies' difficulties in fighting in the horrendous muddy plains evoke an almost surreal portrait of this violence. In even relatively low-urbanized Ireland and Wales, war's impact was not confined to the countryside. It worsened conditions in small Irish towns, where marauding Scots, ostensibly seeking alliance with the colonized Irish, destroyed much more than English property.[36] Rebellion in Wales left the tiny towns there so devastated that it was difficult for them to recover.[37] Towns and their rural hinterlands (*GF*, 133), even in peacetime, sometimes had difficult economic, political and social relations, but they were certainly not always at loggerheads.[38] In wartime, however, the desire for the recovery of urban life often impelled townsmen to expropriate resources from the countryside by violence. The area around Berwick on the Anglo-Scottish borderlands, beset by sieges, reveals the pattern (*GF*, 174–76; F&P, 15).

---

33 This is speculative in the absence of good price data, but on the place of wood in weekly urban markets, see Borgemeister, "Bedeutung des Waldes," p. 24.

34 Thomas Hill, "Wovon lebte die Stadt? Bremens Aussenhandel im Mittelalter," *Niedersächsisches Jahrbuch für Landesgeschichte* 78 (2006), 43–46.

35 Sverre Bagge, "Arms and Means in the Inter-Nordic Conflicts, 1302–1319," *Scandinavian Journal of History* 32 (2007), 5–37.

36 Clarke, "Decolonization," p. 171.

37 Kenneth Murphy, "The Rise and Fall of the Medieval Town in Wales," in *Towns in Decline*, p. 194.

38 Cf. Dyer, "Urbanizing of Staffordshire," p. 25. On determining the extent of urban hinterlands, see besides the reference provided in the text, Hill, "Wovon lebte die Stadt?," pp. 36–38 and 40–43.

The picture is all the more unsettling in that the agricultural sector despite the emerging *blocage* had, by 1300, a long period of vigorous growth behind it (F&P, 13). Recent synthetic and regional work on rural life by Alain Derville, Christopher Dyer, and Ludolf Kuchenbuch has provided more abundant evidence of this fact.[39] Whatever measure one uses, the story is the same: more markets and fairs, more roads and bridges, improvements to navigation on rivers, greater monetization, and augmentations in the number and regularity of commercial exchanges among rural and urban markets (*GF*, 43–48). Even regions long believed to have been immune from some of these developments, like the so-called frozen north (Scandinavia), are now appearing to be much affected (F&P, 14). Svein Gullbekk, for example, has recently demonstrated that Norway enjoyed a far more monetized economy than Kåre Lunden, long the leading economic historian of pre-modern Norway, thought, although one must not exaggerate the level of economic integration which was still low by continental, say, Low Country, standards.[40] The Famine seems to have been the first great shock leading to the temporary reversal of the twin tendencies toward integration and heightened per capita income levels in the north, reflected, I would suggest, in an increase in indebtedness and in the expansion, possibly because of commendation into quasi-serfdom, of labor services on big estates, for example, in Denmark (cf. *GF*, 106).[41] It would probably not be an exaggeration to call the collapse of recently established markets in the north and the declining rate of the creation of new markets in the period an implosion.[42]

---

[39] Alain Derville, *L'Economie française au moyen âge* (Paris, 1995); see also his *Agriculture du Nord au moyen âge (Artois, Cambrésis, Flandre wallonne)* (Villeneuve d'Ascq, 1999). Christopher Dyer, *Making a Living in the Middle Ages: The People of Britain, 850–1520* (New Haven and London, 2002), is excellent, but I demur from the implication (pp. 381–82) that I "play down" the Great Famine in my book; his overall discussion of the famine is on pp. 228–35. Ludolf Kuchenbuch's overview of agricultural developments is interestingly idiosyncratic and more succinct than both Derville's and Dyer's: "Vom Dienst zum Zins? Bemerkungen über agrarische Transformationen in Europa vom späteren 11. zum beginnenden 14. Jahrhundert," *Zeitschrift für Agrargeschichte und Agrarsoziologie* 51 (2003), 11–29.

[40] Svein Gullbekk, "Natural and Money Economy in Medieval Norway," *Scandinavian Journal of History* 30 (2005), 3–19; and Franzén and Söderberg, "Svenska spannmålspriser," pp. 189–214.

[41] Nils Hybel, "Middelalderlig godsadministration i Danmark," *Historisk Tidsskrift* (Denmark) 103 (2003), 268–98 (with English summary).

[42] Samantha Letters, "Markets and Fairs in Medieval England: A New Resource," *Thirteenth-Century England* 9 (2003), 209–23.

It is easier to document regions of high economic integration. Even in areas of low density of population, where barter remained important, however, the production crisis disturbed economic relations more profoundly than one might at first expect. For typically in these areas, per capita production was high, but transportation costs made it uneconomical to export produce and raw materials, excepting luxury goods (furs and amber, for example). Bad weather might reduce production of locally consumed goods, but ordinarily not enough to affect exchange rates. In other words, land or tools or lumber purchased for x bushels of grain in one year could be purchased for x bushels in the next year, in the year after that, and so on. How severe was the production crisis in the Famine years in those regions of Norway where barter was still important? Gunnar Pettersen has shown that in one year from 1315 to 1316 prices for land increased 250%. That is to say, two and one-half times as much grain was needed in 1316 than in a typical non-Famine year to purchase the same plot of land.[43]

Rights in common were already under stress by the increase in population by 1300 (*GF*, 94). The years of the Famine made access to common rights all the more important. The problem is that legal understandings of common rights were not fixed in this period. Two hundred years later a new tenurial regime had vested ownership of the commons (woodlands, waste, pasture, fisheries, and so on) in lords, a development that would subsequently lead to the codification of copyhold tenure, enclosure, rural depopulation, and an explosion in the size of northern towns.[44] Already in the thirteenth century many lords were claiming ownership and insisting that tenants' customary access to the commons did not create an irrevocable right to access. There was, however, far from a consensus in this period on the matter either among lords, between lords and tenants, or among judges (representing state power).

Most claimants of rights (of ownership or customary access) were thus obliged to search the long experiential record, written and oral, and negotiate conflicting claims. It may well be true that the thrust of developing 'clarification' was already favoring lords as a class. This is at least one implication of the phrase 'pressure on common rights,' as it is

[43] Gunnar Pettersen, "Handel og priser i Norge i middelalderen," *Collegium medievale* 13 (2000), 226.

[44] Most standard histories of European agriculture describe these developments. I cannot provide a full treatment here.

used to characterize the period around 1300. Increasingly what lords and other claimants of superior authority in common lands were willing to concede *de facto* was less than what tenants wanted and got. Bad weather, of the kind that was affecting Europe in 1315–22, did not help rectify the situation; it put resolution on hold. An agreement in Suffolk, recorded in the Pakenham cartulary, is revealing in this regard. Clearly, there had been a retreat among many tenants from exercising customary rights in the common against those who claimed superior authority. The schematic might go like this: X agrees not to seek common of pasture in the pasture over which Y claims superior right, if Y will not seek common of woodlands in the woodlands over which X claims superior right. Presumably X and Y had sufficient resources of the two kinds of land in normal times. The years 1315–22 were not normal. The agreement recorded in the Pakenham cartulary and referred to above (dated 1315–20) stipulated that the two contracting parties could not exercise rights of common in each other's lands "unless in manifestly foul weather" (*nisi intempere aperto*).[45]

* * *

One of the most difficult problems facing historians of the early fourteenth century is the assessment of the impact of the Great Famine in relation to the Black Death and the plague cycle. The classical approach has been to go beyond the fact that the Famine killed large numbers of northern Europeans, say, from 5%–10% of the urban population and a somewhat lesser percentage of the rural population (but producing more deaths than that which could be expected in normal times). The argument has been that the seven years of severe malnutrition affecting the poor and the various levels of mild malnutrition affecting more middling sorts of people over such an extended period had a long term effect on the population, especially children, so that as adults they were not as resistant to pathogens. Or, as we would now say, the immune systems of children were negatively affected in 1315–22 (*GF*, 185–87). As adults, in 1348–52 when the plague reached the north, they paid for this immune deficiency with their lives, with perhaps up to 60% of the population perishing.[46]

---

[45] *Pakenham Cartulary*, p. 102, no. 170. Although my hypothetical parallel specifies pasture and woodlands, the agreement itself is silent on this matter.

[46] For the 60% figure, though not for the theory of causation, see Ole Benedictow, *The Black Death, 1346–1353* (Woodbridge, 2004), pp. 273–383.

Two significant problems undermine this scenario. First there does not appear to be any close correlation on a gross aggregate level between high mortality in the Black Death and the extended presence of the Great Famine thirty years earlier. Death rates were comparable in southern Europe and North Africa, where no more than a few brief and localized famines had occurred in the first half of the fourteenth century.[47] Second, a research strategy that I devised and applied to the Famine-affected regions, which might have revealed slight differentiations in death rates, turns out to be of little or no value. It was long thought that at least two northern regions – Bohemia and part of Poland – and perhaps a few others were spared the first attack of the plague. If it could be shown that they also suffered less intense Famine conditions, then one might argue that the theory of immune deficiency among the children in the Famine years played at least a non-trivial contributory role to the high mortality (*GF*, 185–87). Thus, even if aggregate figures, based on huge regions (northern Europe, southern Europe, North Africa), militated against overstressing this factor, the factor could still be real. The problem is that the most recent research, published since the appearance of my study, appears to have demonstrated that the chimera is the widely repeated, though inaccurate, opinion that those two northern regions were spared the plague. New and/or reanalyzed evidence suggests the opposite.[48]

The emerging opinion therefore is that the Great Famine culled the weaker and less productive parts of the northern European population. The decline in population and the rise in per-worker productivity by the death of less productive laborers and the withdrawal of marginal lands from cultivation should have combined – and the evidence is that, with exceptions and for very specific reasons (*GF*, 106),[49] it did combine – to allow the economy to recover fairly quickly. Favorable natural conditions prevailed for nearly two generations under constant technologies and traditional political configurations. Population, distributed in roughly the same way as it had long been distributed,

---

[47] Stuart Borsch, *The Black Death in Egypt and England: A Comparative Study* (Austin, Texas, 2005), pp. 24–25; and Benedictow, *Black Death*, pp. 273–383.

[48] Benedictow, *Black Death*, pp. 216–24. The still unpublished work of Philip Slavin (Economic Growth Center, Yale University) will, however, address the issues discussed in this paragraph in greater depth and with greater sophistication.

[49] William Chester Jordan, *Europe in the High Middle Ages* (London, 2001), pp. 310–13.

recovered. Not surprisingly, *blocage* reappeared. Consequently, and in hindsight, another crisis like that experienced in the Great Famine necessarily threatened. The exogenous variable was not the same. Thirty years after the end of the Famine, the Black Death turned out to be the new exogenous variable.

Indeed, the Black Death and the plague cycle it initiated were such a shock that recovery, of the sort generally enjoyed in northern Europe in the immediate aftermath of the Great Famine, was not possible. Temporary recoveries from the plague, demographically speaking, were succeeded by more than offsetting declines, so that in most of Europe the nadir of population was not in the immediate aftermath of the first attack of the plague but almost one hundred years later.[50] And when demographic recovery to pre-plague levels finally occurred, between 1600 and 1750, depending on region, there had been or were emerging radically different distributions of people (larger cities, less densely populated countryside), significant changes in rural technologies and cropping, and an extraordinary series of transformations in the patterns of European and world trade.

To this extent, it is, strictly speaking, accurate to say that the Great Famine of northern Europe was neither decisive in nor even much of a contributor to bringing the demographic and economic structures of medieval life to an end. Yet, in its own terms, it remains one of the most important and tragic life experiences that Europeans have ever endured.

## *Bibliography*

Bagge, Sverre. "Arms and Means in the Inter-Nordic Conflicts, 1302–1319," *Scandinavian Journal of History* 32 (2007), 5–37.

Barron, Caroline. *London in the Later Middle Ages: Government and People, 1200–1500* (Oxford, 2004).

Benedictow, Ole. *The Black Death, 1346–1353* (Woodbridge, 2004).

*The Bolton Priory Computus, 1286–1325*, ed. Ian Kershaw and David Smith (Woodbridge, UK, 2000).

Borgemeister, Bettina. "Die Bedeutung des Waldes für die Stadt," *Niedersächsisches Jahrbuch für Landesgeschichte* 78 (2006), 17–27.

Borsch, Stuart. *The Black Death in Egypt and England: A Comparative Study* (Austin, TX, 2005).

Carpentier, Elizabeth, and Michel Le Mené. *La France du XIe au XVe siècle: Population, société, économie* (Paris, 1996).

---

[50] Jordan, *Europe in the High Middle Ages*, pp. 296–97.

"The Chronicle of John Somer, OFM," ed. Jeremy Catto and Linne Mooney, in *Chronology, Conquest and Conflict in Medieval England*, Camden Miscellany, 5th series, 10 (Cambridge, 1997), pp. 197–285.

Clarke, Howard. "Decolonization and the Dynamics of Urban Decline in Ireland, 1300–1500," in *Towns in Decline AD 100–1600*, ed. T. R. Slater (Aldershot, UK, 2000), pp. 157–92.

Cohn, Samuel, Jr. *Lust for Liberty: The Politics of Social Revolt in Medieval Europe, 1200–1425, Italy, France, and Flanders* (Cambridge, MA, 2006).

*Court Roll of the Manor of Downham 1310–1327*, ed. M. Clare Coleman (Cambridge, 1996).

Day, W. R., Jr. "The Population of Florence before the Black Death: Survey and Synthesis," *Journal of Medieval History* 28 (2002), 93–129.

Derville, Alain. *L'Economie française au moyen âge* (Paris, 1995).

——. *Saint-Omer des origines au début du XIVe siècle* (Lille, 1995).

——. *Agriculture du Nord au moyen âge (Artois, Cambrésis, Flandre wallonne)* (Villeneuve d'Ascq, 1999).

Dunsford, H. M., and S. J. Harris. "Colonization of the Wasteland in County Durham, 1100–1400," *Economic History Review* 56 (2003), 34–56.

Dyer, Christopher. *Making a Living in the Middle Ages: The People of Britain, 850–1520* (New Haven and London, 2002).

——. "The Urbanizing of Staffordshire: The First Phases," *Staffordshire Studies* 14 (2002), 1–31.

Farmer, Sharon. *Surviving Poverty in Medieval Paris: Gender, Ideology, and the Daily Lives of the* Poor (Ithaca, 2002).

Franzén, Bo, and Johan Söderberg. "Svenska spannmålspriser under medeltiden i ett europeiskt perspektiv," *Historisk Tidskrift* (Sweden) 126 (2006), 186–214.

Gullbekk, Svein. "Natural and Money Economy in Medieval Norway," *Scandinavian Journal of History* 30 (2005), 3–19.

Hill, Thomas. "Wovon lebte die Stadt? Bremens Aussenhandel im Mittelalter," *Niedersächsisches Jahrbuch für Landesgeschichte* 78 (2006): 29–46.

Holt, Richard. "What If the Sea Were Different? Urbanization in Medieval Norway," in *Rodney Hilton's Middle Ages: An Exploration of Historical Themes*, ed. Christopher Dyer, Peter Coss and Chris Wickham (Oxford, 2007), pp. 132–47.

Hurst, J. D. "Fuel Supply and the Medieval Salt Industry in Droitwich," *Transactions of the Worcestershire Archaeological Society*, 3rd series, 19 (2004), 111–32.

Hybel, Nils. "Klima og hungersnød i middelalderen," *Historisk Tidsskrift* (Denmark) 102 (2002), 265–81.

——. "Middelalderlig godsadministration i Danmark," *Historisk Tidsskrift* (Denmark) 103 (2003), 268–98

Jordan, William Chester. *The Great Famine: Northern Europe in the Early Fourteenth Century* (Princeton, 1996).

——. *Europe in the High Middle Ages* (London, 2001).

——. "Famine and Popular Resistance: Northern Europe, 1315–1322," in *Power, Violence and Mass Death in Pre-Modern and Modern Times*, ed. Joseph Canning, Harmut Lehmann and Jay Winter (Aldershot, UK, 2004), pp. 13–24.

——. "Charms to Ward Off Sheep and Pig Murrain," in *Medieval Christianity in Practice*, ed. Miri Rubin (Princeton and Oxford, 2009), pp. 67–75.

Kershaw, Ian. "The Great Famine and Agrarian Crisis in England (1315–1322)," *Past and Present* 59 (1973), 3–50.

Kuchenbuch. Ludolf. "Vom Dienst zum Zins? Bemerkungen über agrarische Transformationen in Europa vom späteren 11. zum beginnenden 14. Jahrhundert," *Zeitschrift für Agrargeschichte und Agrarsoziologie* 51 (2003), 11–29.

Letters, Samantha. "Markets and Fairs in Medieval England: A New Resource," *Thirteenth-Century England* 9 (2003), 209–23.

Murphy, Kenneth. "The Rise and Fall of the Medieval Town in Wales," in *Towns in Decline AD 100–1600*, ed. T. R. Slater (Aldershot, 2000), pp. 193–213.
Nicholas, David. *Urban Europe, 1100–1700* (New York, 2003).
Notte, Ludovic. "La Dévotion à saint Eloi dans les écuries princières (XIIIe–XVIe siècle)," *Revue belge de philologie et d'histoire* 81 (2003), 1051–74.
Page, Mark. "The Peasant Land Market on the Bishop of Winchester's Manor of Farnham, 1263–1349," *Surrey Archaeological Collections* 90 (2003), 163–79.
*Pakenham Cartulary for the Manor of Ixworth Thorpe, Suffolk, c. 1250–c. 1320*, ed. S. D. Church (Woodbridge, 2001).
Pettersen, Gunnar. "Handel og priser i Norge i middelalderen," *Collegium medievale* 13 (2000), 203–28.
Pleij, Herman. *Dreaming of Cockaigne: Medieval Fantasies of the Perfect Life*, trans. Diane Webb (New York, 2001).
Razi, Zvi, and Richard Smith. "The Origins of the English Manorial Court Rolls as a Written Record: A Puzzle," in *Medieval Society and the Manor Court*, ed. Zvi Razi and Richard Smith (New York, 1996), pp. 36–68.
Rippon, Stephen. *The Transformation of Coastal Wetlands: Exploitation and Management of Marshland Landscapes in North West Europe during the Roman and Medieval Periods* (Oxford, 2000).
Rösener, Werner. "Der Wald als Wirtschaftsfaktor und Konfliktfeld in der Gesellschaft des Hoch- und Spätmittelalters," *Zeitschrift für Agrargeschichte und Agrarsoziologie* 55 (2007), 14–31.
Stamm, Volker. "Gab es eine bäuerliche Landflucht im Hochmittelalter? Land-Stadt-Bewegungen als Auflösungsfaktor der klassischen Grundherrschaft," *Historische Zeitschrift* 276 (2003), 305–22.
Stern, Derek. *A Hertfordshire Demesne of Westminster Abbey: Profits, Productivity and Weather*, ed. Christopher Thornton (Hatfield, UK, 2000).
Stone, David. *Decision-Making in Medieval Agriculture* (Oxford, 2005).
Swanson, Robert. *Religion and Devotion in Europe, c.1215–c.1515* (Cambridge, 1995).
Thompson, Augustine. *Cities of God: The Religion of the Italian Communes, 1125–1325* (University Park, Pennsylvania, 2005).
Wendehorst, Alfred. *Das Bistum Eichstätt, I: Die Bischofsreihe bis 1535* (Berlin and New York, 2006).
*Der Wolfenbüttler "Rapularius"*, ed. Hildegund Hölzel-Ruggin, MGH Quellen zur Geistesgeschichte des Mittelalters 17 (Hannover, 2002).

# RABBITS SWIMMING ACROSS BORDERS: MICRO-ENVIRONMENTAL INFRASTRUCTURES AND MACRO-ENVIRONMENTAL CHANGE IN EARLY MODERN HOLLAND*

Petra J. E. M. van Dam

By the end of the Middle Ages rabbits had gained a strong foothold on the elevated, sandy territories of Northern Europe. This comprised the coastal dunes of the county of Holland, situated along the North Sea.[1] These dunes consisted of long ridges of rather dry, sandy hills, interspersed by moist and fertile valleys and plains. The rabbits built burrows in the hills and went out to the plains for grazing. The rabbits were feral. They were introduced by humans in the thirteenth century for the purpose of meat and fur production, but lived in a semi-wild state. Every autumn the rabbits were culled and tens of thousands were sold at urban markets. Sufficient numbers of mature does were left behind, however, to produce next year's harvest, together with a minimal number of bucks to assist them. The people who operated this commercial rabbit culture were called dune lessees (*duinmeiers*), who leased the rights to exploit the rabbits from the count of Holland. For this purpose, the dunes were divided up into rabbit management units called warrens. The dune lessees helped the rabbits to survive the harsh winters by providing fodder and by keeping predators such as foxes and weasels at low numbers. The lessees also dug tunnels for the rabbits to ease burrow building.[2]

---

* I thank Bert van der Valk, Hans Renes and Theo Spek for their comments; and Marjolijn de Vries and Scott G. Bruce for their editorial and linguistic assistance.

[1] In the medieval period Holland was the county of the counts of Holland. During the period of the Dutch Republic (1581–1795) one discerned two regions, the Noorderkwartier and the Zuiderkwartier. Holland is now part of the Kingdom of the Netherlands, and consists of two provinces, Noord-Holland and Zuid-Holland, which have roughly the same borders as the former regions.

[2] My first historical inquiry on the rabbit (*Oryctolagus cuniculus*) in the dunes and the particularities of its 'culture' as a feral animal (introduction, human manipulation of its ecosystem, feeding, selective culling) have been published in: "De rol van de

Other humans used the dunes too, together with other animals. Peasants exploited the valleys for grazing. Large herds of cattle and horses, and some pigs and sheep, lived in the dunes in the summer. In the winter the animals returned to their stables. Cattle and rabbits co-inhabited the flat areas peacefully and, surprisingly, even to their mutual benefit. Cows ate the large tufts of grass, herbs, tree saplings, and even the lower branches of trees. They acted as lawn mowers, a habit which prevented any major forest growth. Rabbits, especially young ones, fed on the tender young grass and herbs that grew up after the cattle had passed. From the early seventeenth century a fascinating description of grazing practices is preserved in a complaint filed by local farmers to the court of Holland in a conflict about grazing rights with the local owner of the dunes, the lord of Aerdenbergh. It claims that at the flat areas the grass was so tall that the cattle had to remove that grass before the rabbits could eat. The argument thus shows that the farmers were quite aware of ecological relationships in their local natural environment.[3] Recent ecological research, carried out in order to improve nature conservation and biodiversity practices, provides systematic evidence for this mutually beneficial relationship. A recent Ph.D. research project about grazing experiments shows that fields grazed by cows sustain more rabbits than those without cows.[4]

---

warande. Geschiedenis en inburgering van het konijn," *Jaarboek Ecologische Geschiedenis 2000* 5 (2002), 59–84. A revised summary of this article was published as "Ein Neubürger in Nordeuropa: Menschliche und natürliche Einflüsse auf die Assimilierung des Kaninchens in den Niederländischen Dünen 1300–1700," in *Beiträge zum Göttinger Umwelthistorischen Kolloquium 2004–2006*, ed. Bernd Herrmann (Göttingen: Universitätsverlag Göttingen, 2007), pp. 167–80. The present article is based on new research into the effect of long-term landscape change on the relationship between rabbits and humans. There is no systematic historical research on grazing or other agricultural uses of the dunes according to Albert J. Thurkow, "Het agrarisch landgebruik in de Hollandse duinen in historisch perspectief," *Duin: Contactblad van Stichting Duinbehoud en Waterwinning* 3 (1986), 75–78. I am currently working on a book about humans and animals in the dunes, 1300–1700, to fill in part of this lacuna.

[3] '... behalven dat dat oock de voors. duynen vlack zijn op veele plaetsen, alwaer veel lang gras is wassende 't welck tot voedtsel van de konijnen onbequaem [is] soo langhe 't selve by de beesten van te vooren niet en is afgheweydt', "Rekwest aan de Edele Mogende Heeren de Staten van Holland en Friesland over het weiden van de beesten in de Aerdenberghse duinen," in *Generale Privilegiën van Kennemerland etc.*, ed. B. van Santen (The Hague, 1665), p. 98.

[4] Interview with plantecologist Elisabeth S. Bakker, "Konijn blijkt dol op koeien (Rabbits love cows)," *Volkskrant*, 20 September 2003, Wetenschapskatern, based on her Ph.D. thesis: *Herbivores as Mediators of Their Environment: The Impact of Large and Small Species on Vegetation Dynamics* (Wageningen: Universiteit Wageningen, 2003), p. 156 and fig. 8.4 B; and Jasja J. A. Dekker, *Rabbits, Refuges and Resources: How Foraging of Herbivores is Affected by Living in Burrows* (Wageningen, 2007), p. 79.

This seemingly Arcadian situation was rudely disturbed in the sixteenth century, when agriculture expanded and new uses of the land came into being in response to market forces. As a result, new types of humans invaded the dunes, humans with holiday homes and extensive gardens where they grew vegetables and fruits. These humans did not want to share the space with rabbits. The rabbits, however, were not very sensitive to the wishes of the new inhabitants. In fact, they kept visiting the plains and were very happy to discover new foodstuffs. The scene was set for centuries of tensions between humans and rabbits, or rather between humans who supported rabbits and those who did not.

Principally, the rabbit interest group had the law on its side. It was a curious feature of medieval hunting legislation that rabbits had the right to graze outside the warrens. Whether the land was used for growing grass, crops or otherwise did not change the matter. But over time a consciousness of fair and unfair grazing rights developed. The acceptance of practices aimed at keeping rabbits off one's premises increased. This led to the building of specific rabbit fences set between the warrens and adjoining grounds. Rabbit fences came in all sizes and shapes, varying from large wooden structures to simple ditches filled with water.[5]

In this article I will treat the history of the rabbit fences of the coastal dunes of the Netherlands as a contextual environmental history. The larger framework of the argument is the debate about the nature-culture dichotomy. The first authors who wrote about the relationship between humans and nature, such as the famous nineteenth-century American George Perkins Marsh or the Englishman Alfred Wallace, expressed their awareness that there was very little pristine nature left on earth.[6] This consciousness remained important among later writers, especially among those in the United States who called themselves environmental historians, active from the 1970s onwards. An important threshold was crossed when the concept spread more widely among environmental historians that what is nature (or culture) is

[5] There were no fences in between the warrens. The borders between the warrens were indicated by tokens in the terrain, such as trees, characteristic (named) dune tops, roads, and poles erected by the dune lessees.

[6] George Perkins Marsh, *Man and Nature, or Physical Geography as Modified by Human Action* (London, 1864), cited by Roderick Nash, "The State of Environmental History," in *The State of American History*, ed. Herbert J. Bass (Chicago, 1970), p. 251; and Alfred Wallace, *The Wonderful Century, Its Successes and Its Failures* (London and New York, 1898), pp. 367–79.

always in the eye of the beholder. So concepts of nature or culture differ across time, across culture and even, at a smaller scale, among social groups or individuals.

The problem is very clear in those areas that people of our time consider as nature, and the dunes of Holland are such an area. Here we must look more precisely. 'People of our time', in such cases, often means urban people: scholars, nature-lovers, tourists. They regard any landscape with plants and animals as nature. In Europe, however, most of that is agricultural landscape, heavily influenced by thousands of years of manipulation. But forests, dunes and heaths have also been influenced by human agency, although less intensively and thus only visible to the specialist, such as the forester or the ecologist. Burning regimes, hunting, herding and the exploitation of fruits, leaves, dead wood, sod and other biomass have changed such landscapes profoundly over time. Human agency changed the shape of the trees; it changed the plant varieties; and it affected the soil formation, to sum up but a few examples. In a recent paper Emmanuel Kreike introduced the concept "environmental infrastructure" to denote what he calls the growing grey zone between Nature and Culture. It includes "cultivated landscapes," coppice woodland, farms and fields, including fences or demarcated clearings.[7] Kreike created this term to describe human-nature relationships in an African context, but his reflections are valuable for the European landscape as well, in particular for areas that were exploited in an extensive way, such as the dunes of Holland.

Borrowing the term environmental infrastructure contributes to framing the history of animal fencing in the larger picture of environmental change. One can consider including all varieties of means that humans employ to create distance between animals and themselves, to give animals a separate place and to seclude animals in a designated territory. In our case, this primarily included vertical fences made of all sorts of material. But also it included more horizontal means, such as broad ditches with running water, which were particularly appropriate where the dune massive bordered wetlands. All of these we can consider as micro-environmental infrastructures.

---

[7] Emmanuel Kreike, "Architects of Nature: Environmental Infrastructure and the Myth of Natural Resources Management" (paper presented at the Fourth Conference of the European Society of Environmental History, Amsterdam, The Netherlands, June 5–9, 2007).

The practice of building rabbit fences is not confined to Holland. The most famous rabbit fence was built shortly after 1900 and crossed Australia for over two thousands of kilometres. In that case, the fence protected the interests of humans with sheep.[8] Enclosing feral animals such as rabbits in relatively large habitats serves as a kind of prefiguration of our modern nature reserves or zoos. These examples connect the history of rabbit fencing in Holland to our modern experience.

In this article I will investigate what happened when new needs for separating humans and rabbits arose in the coastal dunes of the Netherlands between 1400 and 1700. How did the history of fences develop? How did human and natural agency interact with respect to micro-environmental infrastructures? I will compare two districts of the Netherlands, Holland north and south of the marine inlet known as the IJ. This comparison gives access to multiple sets of explanations, taking into account landscape changes, institutional frameworks and economic developments.

## *The Dutch Coastal Dunes*

At the end of the Middle Ages the coastal dunes of Holland were a relatively new landscape. In the Roman Period, the coast had stretched further seawards and the dunes were much lower, rising only a couple of meters above sea level. Starting in the ninth century, the sea attacked the coast and destroyed the first rows of dunes. All the sand loosened by the sea was taken inland by the prevailing western winds. Waves of sand dunes moved over the old dune landscape. Gradually the masses of sand stabilized again and all sorts of pioneer plant communities sprang up. Soil formation took place and more demanding plants moved in. By 1300, directly along the North Sea beach, the coastal landscape consisted of alternating rows of dunes, up to probably 40 meters in height, interchanging with so-called dune valleys. Geologists call these dunes the Younger Dunes. Going inland, the dunes were lower and the flat spaces, called beach plains, broader. The lower, inland dunes

[8] John McNeill, *Something New Under the Sun: An Environmental History of the Twentieth Century* (London and New York, 2000), p. 254. I found some evidence for sheep raising in the dunes, which deserves future attention, for sheep, unlike cattle and pigs, are rivals to rabbits, of which the situation in Australia has provided abundant proof. In literature, the ratio of fodder consumption between these animals varies from 4 to 10 rabbits per sheep.

were in fact remnants of the Old Dunes of the period before 1000.[9] The coastal dunes of Holland form a chain of stable dunes along the North Sea coast that now runs from the navy town of Den Helder to the large commercial sea port of Rotterdam. At the end of the Middle Ages, the chain of dunes was shorter, because in the north coastal formation was still taking place (the medieval islands 't Oge and Huisduinen are now part of the continental chain of dunes) and the chain of stable dunes started at the village of Petten.[10]

Holland consisted of two geographic regions, Holland north and south of the IJ. Holland north of the IJ was almost a peninsula, situated between the North Sea in the west and a big inlet of the North Sea, the Zuiderzee, in the east. At its foot a small inlet of the Zuiderzee, the IJ, protruded deeply into the land up to the foot of the dunes. In Holland north of the IJ, the seas remained very invasive, in particular on the occasion of storm surges. The land was covered with large and ever growing lakes.[11] Holland south of the IJ was much better protected against the sea by its high-lying sweet-water peat bogs and the land-masses of the central Netherlands, such as the region of Utrecht. Here the Old Dunes came in clusters of hills, interspersed by large, low-lying beach plains. In earlier (pre-)historical periods, these plains had been beaches, for the coast had had a very turbulent history with periods of advancing and retreating seas. Over time, these so-called beach plains had become covered with bogs consisting of lots of water-loving plants,

---

[9] Saskia Jelgersma et al., "The Coastal Dunes of the Netherlands: Geology, Vegetational History and Archaeology," *Mededelingen Rijksgeologische Dienst* n.s. 21 (1970), 93–167; Rob Rentenaar, "De Nederlandse duinen in de Middeleeuwse bronnen tot omstreeks 1300," *Geologisch Tijdschrift* 11 (1977), 369; Waldo H. Zagwijn, "The Formation of the Younger Dunes on the West Coast of the Netherlands (AD 1000–1600)," *Geologie en Mijnbouw* 63 (1984), 259–68; *Man-Made Lowlands: History of Water Management and Land Reclamation in the Netherlands*, ed. Gerardus P. van de Ven (Utrecht, 1993), pp. 22 and 24; Willem A. Ligtendag, "De kustlijn in kaart," *Geografisch tijdschrift* 5 (1990), 380–7; Henk Schoorl, *De Convexe Kustboog. Texel-Vlieland-Terschelling 1: Het westelijke Waddengebied en het eiland Texel tot circa 1550* (Schoorl, 1999), pp. 14–30; and M. van Koningsveld et al., "Living with Sea-Level Rise and Climate Change: A Case Study of the Netherlands," *Journal of Coastal Research* 24 (2008), 395–407.

[10] Henk Schoorl, "Wording en gebruik van het Nederlandse duinlandschap," in *Kust en kaart: Artikelen over het kaartbeeld van het Noordhollandse kustgebied* (Schoorl, 1990), p. 106.

[11] The dunes of Holland were still one uninterrupted strip in the Middle Ages. Only in the nineteenth century was a channel dug linking the IJ directly to the North Sea, following the construction of the North Sea canal, with elaborate harbour defences and locks at IJmuiden. This is still the current situation.

such as reeds and mosses. The drainage had always been very bad because of all the surrounding higher grounds, so peat formation had taken place.[12] By the end of the Middle Ages, these beach plains were covered by peat layers of up to 2 meters deep.

Information about the creation of this specific coastal landscape comes primarily from geologists and archaeologists and it is supported by incidental written sources relating to periods of intensive sand drift. Only in the sixteenth century do maps of the coastal area begin to appear. For our mental picture of the medieval dunes these first maps are very influential. Basically, they show rows of high dunes and small dune valleys along the coast, low dunes further inland, and the large, flat beach plains in between. The one feature that stands out is that this was a 'linear' landscape. The long, alternating ridges of dunes and beach plains were formed in north-south parallel rows.[13]

The early seventeenth-century survey maps such as those drawn by father and son Van Berckenrode (1610–1615) show where humans had settled. The oldest settlements were situated at the high parts of the Old Dunes, safe and dry places, where fields for agriculture could be laid out. The villages were situated in rows, following the dune ridges, with long stretching north-south roads connecting the settlements. At the seaside specialized fishing villages were founded, which connected to the existing north-south structure by a network of west-east roads. The villages at the Old Dunes were the oldest, some tracing their origins back to Merovingian times. Most fishing villages probably started after the period of the formation of the Younger Dunes. From 1300 onwards, a new phase of reclamation started in the dune region, directed at draining and reclaiming the peaty beach plains. Humans dug long north-south canals for drainage, probably connecting and canalizing

---

[12] This process of peat formation is better known for the massive inland peat bogs. Almost no traces of peat are left in the dune area but the reclamation pattern in the current landscape and the many incidences of peat digging offenses in the sixteenth century preserved in the archives of the Regional Water Board of Rijnland are strong pieces of evidence for this industry, but they have not been fully investigated: Jan J. J. M. Beenakker, *Lisse op de grens van droog en nat: De bewoningsgeschiedenis en landschapsontwikkeling van een dorp in de bloembollenstreek tot omstreeks 1900* (Lisse, 1993), p. 41; and Leiden, hoogheemraadschap van Rijnland, Oud-Archief [hereafter, OAR] inv. no. 14 and 15. Recent studies have revealed a similar landscape history for coastal Flanders as well: Beatrijs Augustyn, *Zeespiegelrijzing, transgressiefasen en stormvloeden in maritiem Vlaanderen tot het einde van de XVIde eeuw: Een landschappelijke, ecologische en klimatologische studie in historisch perspectief*, 2 vols. (Brussel, 1992).

[13] OAR, map collection, A1/A101 (map of dunes and fields near The Hague, drawn by Coenraat Oelenz, 1555).

small, natural creeks. These activities reinforced the linear, north-south spatial pattern.[14]

In this article I will concentrate on the large centrepiece of the dune masses that ran from the village of Egmond to The Hague. In this territory the Count of Holland held most of the rights to rabbit culture and the administration of this survives fairly completely. In these dunes two parts can be distinguished. The part north of the IJ ran from Egmond to somewhat below the IJ, the part south of the IJ ran from the latter to The Hague. The IJ was a major geographic divide and a watershed. Roughly, south of this watershed the Regional Water Authorities of Rijnland coordinated water drainage at the regional level. North of the IJ no such regional authorities existed.

It is difficult to give the exact measures of the dunes at a particular time, since their total surface area underwent a lot of changes over the centuries, due to both natural and human agency. To give an indication, one can use the maps from the early seventeenth century. The territory included in this case-study was some 65 kilometres long as the crow flies. Its width varied from about one to three kilometres, measuring from the North Sea beach to the outer border of the Old Dunes. The sizes of the warrens varied greatly, from hundreds to thousands of square meters.

### *Unreasonable Grazing*

From a legal point of view, the warren rights in the Dutch dunes were part of the hunting laws. Formally, feeding and protecting the rabbits in warrens was simply meant to support rabbit hunting, as J. G. G. Jelles explains in his history of the dunes of Kennemerland.[15] A special development of the hunting laws, however, was the right of *afweyding* (offgrazing). This was grazing beyond the borders of the warrens, thus beyond the so-called wilderness, on the grounds of private landowners. The important presumption was that it was forbidden for landowners

---

[14] *Prins Maurits' kaart van Rijnland en omliggend gebied door Floris Balthasar en zijn zoon Balthasar Florisz van Berckenrode in 1614 getekend*, ed. Kees Zandvliet (Alphen, 1989), nos. 4, 8, and 12 (1614); and the map of the dunes between Wijk aan Zee and Egmond drawn by G. T. Langedijk in 1614 and published in J. G. G. Jelles, *Geschiedenis van beheer en gebruik van het Noordhollands Duinreservaat* (Arnhem, 1968), Appendix 2.

[15] Jelles, *Geschiedenis*, p. 44; and Johan Ph. de Monté Ver Loren and Johannes E. Spruit, *Hoofdlijnen uit de ontwikkeling der rechterlijke organisatie in de Noordelijke Nederlanden tot de Bataafse omwenteling* (Deventer, 1982), p. 130.

to kill the rabbits invading their fields coming from the wilderness, even if they damaged crops. Even outside the wilderness, game remained the property of the lord, the owner of the wilderness rights.[16]

Wilderness is a concept affiliated to medieval rights called *regalia*. The count of Holland owned the right to all wild, uninhabited territory, including peat bogs, dunes and all new lands along rivers and the sea, which were formed by natural processes of sedimentation. The concept of wilderness originated in Roman times and was kept alive by legal specialists at the courts of European princes.[17] The counts of Holland had maintained this right as well. It does not mean, however, that the wilderness of the Dutch dunes were pristine as, for instance, in the modern American sense of wilderness. The dunes had been inhabited for thousands of years, and certainly long before the counts of Holland arrived in the tenth century A.D. People lived, as it were, on islands of civilization consisting of villages and their adjacent fields. The villagers would use the surrounding 'nature' for herding, collecting fuel and fruits, and so on. From the count's perspective this was his wilderness and he founded his warrens there. Thus a potential for conflicts arose. All such rights were user rights, however, not territorial rights, so conflicts were often avoidable. Practical solutions were found, depending on the local power constellations. In the dunes, the villagers could hire grazing rights from the count, to the effect that thousands of their cattle and horses and some sheep and swine roamed in the wilderness as the detailed fourteenth-century domanial accounts of the count show. Other rights for gathering fuel wood and sods were for hire, too.[18] Thus the so-called wilderness was full of extensive forms of land use, in particular animal raising. The count valued claiming and maintaining his right of wilderness, because it gave him substantial rights of income.

The principle of offgrazing is at least as old as and probably even older than the occurrence of game warrens in Holland, the first

---

[16] Originally, the lord had the rights to so-called noble game (deer, swine) and the local free people had the rights to smaller game such as hares. In most of Holland, however, the local people had lost these rights. Legal historians call this the feudalisation of the small game. See Monté Ver Loren, *Hoofdlijnen*, p. 129.

[17] Monté Ver Loren, *Hoofdlijnen*, p. 123.

[18] Dick E. H. de Boer, *Graaf en grafiek. Sociale en economische ontwikkelingen in het middeleeuwse 'Noordholland' tussen 1345 en 1415* (Leiden, 1976), pp. 265–73, gives the monetary incomes from the grazing rights; the figures are mine, based on unpublished research.

reference to which dates from the thirteenth century. The count of Holland gave all rights of warren to his wife in 1297 and this included several types of game, including rabbits.[19] The offgrazing of rabbits was probably not essentially different from the offgrazing of deer or other game living in warrens, but the history regarding rabbits stands out because of the sheer quantity of them and the curious conflicts that arose concerning offgrazing rights, which are also valuable sources of information.

The eighteenth-century scholar and agronomist *avant-la-lettre* Jan Kops (1765–1849) gave a short history of the right of offgrazing as exercised by rabbits in agricultural fields along the dunes in his 1798 treatise *Current State of the Dunes* (*Tegenwoordige staat der duinen*).[20] According to him the first document issued by the central government that describes the right of offgrazing was the ordinance about the sides of the ditches in the dunes by Charles V in 1550. But the first use of the specific expression 'the offgrazing of the rabbits' did not appear in a similar ordinance until 1598.[21] I presume that the description of 1550 refers to a practice that was much older, and that in the sixteenth century an important new attitude towards the old practice grew, which led to more explicit legal definitions such as the invention of the word offgrazing. In this period the central government of the Habsburg emperor took care to lay down and make explicit all of its rights in relation to its subjects, as part of a wider process of state formation and reinforcement of principles of Roman law. A similar movement occurred all over Europe regarding the rights of peasants to graze their livestock in forests.[22] In the process the offgrazing rights may have become more deeply entrenched in legal thought and jurisdiction.

Kops discerns reasonable and unreasonable offgrazing. Unreasonable offgrazing existed when the right covered a surface of three to four times the size of the warren from which the rabbits originated. In his time, in extreme cases dune parcels of two *morgen* had 20 *morgen* of

---

[19] Rob Rentenaar, "De vroegste geschiedenis van het konijn in Holland en Zeeland," *Holland: Regionaal-Historisch Tijdschrift* 10 (1978), 5.

[20] Jan Kops, *Staat van Holland's duinen en ontwerp tot vrugtbaarmaking van dezelve, opgegeven in het rapport der comissie over het onderzoek der duinen van Holland. Tegenwoordige staat der duinen (van het voormalige Holland)*, vol. 1 of 2 vols., 2nd ed. (Utrecht, 1818).

[21] Ordonnance of 6 February 1598, art. 22, published in Paulus G. F. P. N. Merula, *Placaten ende ordonnancien op 't stuck van de wildernissen* (The Hague, 1605), p. 246.

[22] Joachim Radkau, *Nature and Power: A Global History of the Environment* (Cambridge, 2008), p. 138.

offgrazing rights (a *morgen* is 0.85 ha). Kops explains how the right of offgrazing related to rabbit fencing. For him it is a rational choice. Either one allows the offgrazing on one's fields and tolerates diminished harvest incomes or one invests in putting up fences.[23] Either way, the presence of rabbits in the wilderness meant high costs for tillers outside the wilderness. For my argument the main feature to remember is that in the early modern period, rather than allowing the medieval custom of offgrazing, the erection of rabbit fences became the rule. Nobody was allowed to kill rabbits but officials of the count; this medieval principle persisted.

### *Changes in the Land: Drainage*

From the end of the Middle Ages onwards, drastic changes in land use occurred in the dune region, in particular in the more inland Old Dunes. Old activities were revived and intensified, such as peat and sand mining, while new activities sprang up, including horticulture and linen bleaching. Such changes affected the relationship between humans and rabbits profoundly. The new entrepreneurs were highly motivated to build rabbit fencing and they had sufficient capital to invest in this endeavour.

A fundamental prerequisite of new and more intensive land use in the low-lying plains between the dune ridges was the extension of the drainage system. The first documents of this development date from the fourteenth century, the period when the dune villages had just started to expand and reclaim the wetland areas below the settlements (which were situated on the dune tops). Some major waterways were dug, sometimes turning already existing small dune creeks and former sea inlets into human-controlled drainage canals. For instance, in The Hague near the castle of the counts of Holland the so-called Haagse Beek was extended with a canal. It fed the fresh and very clean dune water into the castle pond.[24] Between Haarlem and Leiden the dune ridges were cut through from the one dune beach plain to the next to establish major water drains known from projects executed in 1307

---

[23] Kops, *Tegenwoordige staat*, p. 133.

[24] This building now serves as the national House of Parliament and the pond is still fed by dune water, but its use is mainly restricted to recreational winter ice-skating by the urban populace. See Maarten van Doorn and Jaap Mennema, *De Haagse Beek: een (natuur)historische verkenning* (The Hague, 1992), pp. 9–15 and 30–5.

and 1403.[25] In this way the surplus water of the plains between the dune ridges gained access to a big lake system, which drained into the sea, and which by the sixteenth century would become known as the Great Harlem Lake.[26] A second phase of building major canals along the dunes occurred in the sixteenth and seventeenth centuries. For instance, south of the IJ the dune excavation near Lisse called Mallegat (1590) was essential for the improvement of the drainage of three villages, Noordwijkerhout, Voorhout and Lisse.[27] North of the IJ the new Hoepbeek (1613) drained the low-lying parts of the dune villages of Castricum and Bakkum.[28]

Regarding the micro-drainage system, for the medieval period we can safely assume that in connection with the (documented) major drains, hundreds of (undocumented) small drainage canals were dug in the dune planes. For the later period, conflicts related to the jurisdiction of canals, bridges and sluices, as well as permission for digging through regional watersheds and making connections to the existing network, are systematically preserved in the archives of the Regional Water Authorities, and more incidentally, in the archives of local lords and major landowners, such as abbeys and urban charitable institutions.[29] In addition, the sixteenth century was the period when local drainage was much improved by the proliferation of small, wind-driven drainage mills. Such mills had existed in Holland since 1408, but the early ones were mainly private. The new feature was that the mills were owned publicly by groups of landowners, who erected public water

[25] Sybrandus J. Fockema Andreae, *Het Hoogheemraadschap van Rijnland. Zijn recht en zijn bestuur van den vroegsten tijd tot 1857* (Leiden, 1934, repr. Alpen aan den Rijn, 1982), 68; Beenakker, *Lisse*, p. 44; and Documents about maintenance 1408, OAR, inv. no. 6052.

[26] Petra J. E. M. van Dam, "Sinking Peat Bogs: Environmental Change in Holland, 1350–1550," *Environmental History* 5 (2000), 32–45.

[27] File of documents concerning digging and maintenance of the Mallegat 1588–1615, OAR, inv. no. 5692; map of the dunes near Mallegat, OAR, inv. no. A 506; and Document on widening the canal along the Gravendamseweg (8 april 1595), OAR inv. no. 7807.

[28] Its history is well documented since it happened to be situated in the territory of the leading politician of the Dutch Republic of the period, Johan van Oldenbarnevelt. See Frits D. Zeiler, *Nollen, krochten, blinken: duintoponiemen tussen Wijk aan Zee en Camperduin* (Castricum, 1995), pp. 3 and 5; S. P. A. Zuurbier, "De heerlijkheid Bakkum en zijn ambachtsheerlijkheid," *Oud-Castricum: Jaarboekje* 3 (1980), 10; and Johan W. Groesbeek, *Middeleeuwse kastelen van Noord-Holland: hun bewoners en bewogen geschiedenis* (Rijswijk, 1981), pp. 128–41.

[29] Beenakker, *Lisse*, p. 41; OAR, inv. no. 14, fols. 7v and 58v; OAR, inv. no. 15, fols. 58, 75v, 83 and 127v; OAR, inv. no. 16, fols. 95, 104, 135, 138v; and OAR, inv. no. 4235.

drainage units called 'polders.'[30] Improved drainage was the condition *sine qua non* for the changes in the land. On the one hand, drainage was needed to reclaim the low-lying wetlands, and, on the other hand, new land use yielded profits that could be invested in the water infrastructure. Furthermore, for our main argument about rabbit fencing, it is important to remember that all small and large canals could be integrated into fencing systems, as I shall demonstrate below.

### *Changes in the Land: Peat Mining, Sand Mining, Horticulture*

From 1400 or so, from the inland side the Old Dunes were gradually excavated and leveled. The large stretches of the plains covered by a thin layer of peat were mined. After drying, peat made a good fuel and most of it was transported to the urban markets. In contrast to the deep peat bogs in the interior, where the intensive mining left swamps and lakes from the early modern period onwards, along the coast relatively high sandy plains remained which could be easily drained.[31] Levelled ground was much in demand. The sandy soil was prepared for agriculture using nutritious soils, dug up locally from lower sediments, and using night soil (urban refuge), which was often a return freight for the sand boats, as reclamation contracts indicate.[32]

Large stretches of the sandy dunes were removed through sand mining. Some sand was applied locally for levelling and repairing roads and embankments, but most was sold at the market and served several purposes. Some was used as ballast for ships, yet enormous amounts of sand were shipped off as construction and filling material. In the

---

[30] Milja van Tielhof and Petra J. E. M. van Dam, *Waterstaat in stedenland: De geschiedenis van het hoogheemraadschap van Rijnland voor 1857* (Utrecht, 2006), pp. 134–39; and Siger Zeischka, "Minerva in de polder: waterstaat en techniek in het hoogheemraadschap van Rijnland (1500–1856)," (Ph.D. thesis, Vrije Universiteit, Amsterdam, 2007).

[31] In the Middle Ages, the inland peat bogs were mined down to the groundwater level. From 1510 onwards, dredging for peat became common, which created large bodies of water. See Charles Cornelisse, *Energiemarkten en energiehandel in Holland in de late middeleeuwen* (Hilversum, 2008); Guus J. Borger, "Draining-Digging-Dredging: The Creation of a New Landscape in the Peat Areas of the Low Countries," in *Fens and Bogs in the Netherlands: Vegetation, History, Nutrient Dynamics and Conservation*, ed. Joseph T. A. Verhoeven (Dordrecht, 1992), pp. 131–71; and Van Dam, "Sinking Peat Bogs."

[32] Contra-letter of Nicolaes van der Laen, burgomaster of Haarlem, 11 February 1563, NA, Grafelijkheids Rekenkamer, Charters niet-Suys, doos rood 59 (not inventoried, but indicated by n. 433, m. 20, no. 1285).

fifteenth century, at the confluence of a dune canal and the river Spaarne, an anchoring place existed where ships loaded dune sand. In the sixteenth century, specially designated sand loading places existed in the town of Haarlem.[33] Villages situated in the massive inland peat bogs applied sand 'to keep floating': a farmer would put a layer of sand on the soil before building his house and tilling ground was also raised in this way. Archaeologists have found many traces of such practices.[34] On a grander scale this also happened in the towns that bordered on the peat bogs. Before building extensions into the wetlands, the soil was prepared by contractors who spread a thick layer of sand over the developed site.

The first documented, incidental cases of dune leveling through sand mining are known from the fifteenth and sixteenth centuries and are situated around the city of Haarlem.[35] In this area the first sand canals (*zandvaarten*) were dug made for the transport of the sand to the city. By the end of the sixteenth century, sand mining had become a full-grown business. The great demand from Amsterdam, a city that experienced a rapid growth since the end of the sixteenth century, may have been one of the main reasons for this.[36] Another reason certainly was the rising demand for tilling land.

A well-documented and well-researched case of a sand mining project concerns the *Keukenduin* (Kitchen Dune) at the village of Lisse, between Haarlem and Leiden. In 1604 a consortium acquired permission to exploit a piece of land 275 *morgen* (234 hectares) in size. The main shareholders were noble members of the local village elite, such as Andries van Thienen, Esquire, who occupied the post of regional bailiff, and members of the patriciate of the town of Haarlem, such as Gerard van der Laen, related to the burgomaster of the town, Nicolaes

[33] J. H. M. Sloof, "Rijnland en het Spaarne in de late middeleeuwen," in *De loop van het Spaarne: De geschiedenis van een rivier*, ed. Bert C. Sliggers (Haarlem, 1987), pp. 17–35; and *Rechtsbronnen der stad Haarlem*, ed. Johan Huizinga (The Hague, 1911), pp. 186, 243–45 and 316.

[34] Van Tielhof and Van Dam, *Waterstaat in stedenland*, p. 131.

[35] Pier Hoekstra, *Bloemendaal, proever ener streekgeschiedenis* (Wormerveer, 1947), p. 84; and *Informacie up den staet, facultueyt ende gelegentheyt van de steden ende dorpen van Hollant ende Vrieslant om daernae te reguleren de nyeuwe schiltaele gedaen in jare MDXIV*, ed. Robert Fruin (Leiden, 1866), p. 45.

[36] Boudewijn Bakker, "De zichtbare stad 1578–1813," in *Geschiedenis van Amsterdam: Centrum van de Wereld, 1578–1650*, ed. Maarten Prak and Willem Frijhoff (Amsterdam, 2005), pp. 17–48. Such practices still exist in the low-lying western Netherlands, but other sand resources are being exploited nowadays, such as the bottom of the Northsea.

van der Laen. Most shareholders already had large properties and mansions in Lisse or surrounding villages, so we may assume they had access to the necessary local power and knowledge networks. At first, business was not profitable. The costs for the water infrastructure may have been higher than expected. Starting in 1605, sand workers deepened and widened the main water drain into the Harlem Lake system, the aforementioned Mallegat, and they began working on four sand canals in the plot. Perhaps the price of sand was lower than expected. In any case, shareholders struggling with a lack of sources for investment withdrew and had to be replaced. Finally, in 1635, Adriaen Block Maertensz, a man who had become rich as a shipper for the United East-Indan Company (*Verenigde Oost-Indische Compagnie*), joined the consortium. He built a sumptuous mansion on the levelled dune parcel and added a farm in 1643.[37]

Next to sand mining another new activity began in the sixteenth century on the leveled beach plains: the cultivation of vegetables and fruits. In the late Middle Ages, horticulture had been situated on heavy clay soils, mostly around cities, close to the consumers. Famous examples are the gardens around Leiden, a town built on the banks of the river Rhine. Medieval horticulture had focussed on the so-called heavy vegetables such as beets, carrots, cabbage and peas. In the sixteenth century, however, the cultivation of vegetables not only changed location, but also changed character. New, so-called light vegetables became fashionable among the urban and country elites. These are classified as leaf and fruit vegetables, including artichokes, asparagus, endives, cucumbers, melons, pumpkins, salad, spinach, and several kinds of berries, such as strawberries.[38] The light soils of the beach plains and leveled Old Dunes were very suitable for this new horticulture. So, all along the dunes, farms began specializing in the new plants.

---

[37] Alphons M. Hulkenberg, *Keukenhof*, Hollandse studiën 7 (Dordrecht, 1975), pp. 9–27. The Keukenhof estate has become widely known, because in the twentieth century the national bulbflower showgarden was founded at the property.

[38] W. J. Sangers, *De ontwikkeling van de Nederlandse tuinbouw* (Zwolle, 1952), pp. 114 and 134; Jan de Vries, *The Dutch Rural Economy in the Golden Age, 1500–1700* (London, 1974), p. 154; Jan Bieleman, *Geschiedenis van de landbouw in Nederland 1500–1950: Veranderingen en verscheidenheid* (Meppel, 1992), 72; *De tuin van Holland. Geschiedenis van tuinbouw en groenteveiling in Leiden*, ed. Johannes G. Endhoven et al. (Utrecht, 1992), pp. 9–13; and Johanna M. van Winter, "Green Salads: An Innovation in the Diet of the Renaissance Period," in *Spices and Comfits: Collected Papers on Medieval Food*, ed. Johanna M. van Winter (Totnes, 2007), pp. 369–81. Most of these plants came from Southern Europe, but some came from the New World, such as "winterpostelein" (*Claytonia perfoliata*), which became wild in the dunes.

In the same area, in the seventeenth century the urban elites built mansions where they retreated from the hectic city life in the summer. The mansions had ornamental gardens and amusement parks, but also elaborate fruit and vegetable gardens, and they grew a lot of marketable produce.[39] The difference between mansions and farms was often small, and the one might even grow out of the other. Mentioned above is the case of Maertensz who first built a mansion and added a farm later on. Other farms, founded in the early sixteenth century, were expanded and converted to luxurious mansions in the seventeenth century.[40] Between the urban market and the centres of horticulture that sprang up, very good transport connections were established. For instance, between Beverwijk and Heemstede and the city of Amsterdam market boats operated regularly, which took vegetables to the city and returned with night soil.[41]

Midway through the seventeenth century yet another form of land use became popular. From about 1500 onwards, linen bleaching was established in dune valleys, beach plains and on leveled grounds around the city of Haarlem, which was the centre of linen production and trade. The dunes were very suitable for this refining process, because they contained rich reserves of high quality groundwater. Basically it was rainwater. The other ingredients for the bleaching suds were also easily available. The acid came from buttermilk, a byproduct of making butter. This was available in large quantities, since dairy production was the dominant activity in large parts of Holland north of the IJ and also on the dune plains near Haarlem. Soap and ingredients for soap such as ash were shipped in great quantities from Scandinavia by merchants from Haarlem. By the end of the sixteenth century, forty enterprises that bleached linen cloth were situated in the dunes area considered and the number increased over the seventeenth century, as indirect evidence suggests. By 1650, over 1500 people worked in this business.[42]

[39] Frederik H. Horsten, "Het sociaal-economische leven: Landschap en geografie in het Noorden 1300–1500," *Algemene Geschiedenis der Nederlanden* [hereafter, *AGN*] 2 (1981–1982), 18–39; and J. D. H. Harten, "Het sociaal-economische leven, geografie en demografie 1500–1800: Landschap in beweging," *AGN* 5 (1979–1980), 38–77.

[40] J. D. H. Harten, "Stedelijke invloeden op het Hollandse landschap in de 16e, 17e en 18e eeuw," *Holland: regionaal-historisch tijdschrift* 10 (1978), 114–34; and Erik de Jong, "For Profit and Ornament: The Function and Meaning of the Dutch Garden Art in the Period of William and Mary 1650–1702," in *The Dutch Garden in the Seventeenth century*, ed. John D. Hunt (Washington, D.C., 1990), pp. 24 and 29.

[41] Sangers, *De Nederlandse tuinbouw*, p. 130.

[42] S. C. Regtdoorzee Greup-Roldanus, *Geschiedenis der Haarlemmer bleekerijen* (The Hague, 1936), pp. 9, 13, 21, 28, 102, 105 and 323; and Jan Lucassen, *Naar de kusten van de Noordzee: Trekarbeid in Europees perspektief, 1600–1900* (Gouda, 1984), p. 106.

Bleaching entrepreneurs were among the people interested in developing dunes into commercially useful territory, though presumably on a smaller scale than the business of the commercial sand miners mentioned above. One example of this concerns Lord Barthout of Assendelft, who in 1597 received permission from the Lady of Brederode to extend his property at the border of the villages of Santpoort and Velsen by about one hectare. He dug into the high dunes for water and turned the dune valleys into bleaching grounds, drained by ditches. Of course, neither muddy rabbit prints nor other traces of rabbits would be welcome on the white cloth stretched out on the bleaching fields. So the Lord of Assendelft obtained permission to procure building materials for making fences, including the right to dig up sod (to cover earthen walls) and thorny bushes (to make living fences on top of the walls) in the as yet unreclaimed dunes.[43] Very beautiful seventeenth-century paintings of some of the bleaching enterprises exist, which show the buildings and the fields with the bleaching cloth, interspersed by canals and lined with trees and fences.[44] All of the entrepreneurs engaged in the activities described so far, including peat and sand mining, intensive horticulture, and linen bleaching, had strong incentives to keep rabbits off their lands, and what is more, they had the capital to erect rabbit defence systems.

## *Dry and Wet Fencing*

Animal fences as such were of course not an invention of the sixteenth century.[45] From a long-term perspective, however, a difference in scope came about. We can see this clearly if we reflect on how fencing can function on a broad spectrum. At the one end of the spectrum, all humans and animals live together in one space, without or with only

---

[43] Document concerning the financial arrangement of a parcel of 1 morgen and 1 hont in the dunes of Brederode, produced by Warnar van Batenburg and Herbert Stalpaert van der Wiele on behalf of the Lady of Brederode on 14 August 1597 to Jonkheer Barthout of Assendelft, 27 October 1597 (NA, Grafelijkheidsarchief, Rekenkamer, Registers [hereafter GRRREG] inv. no. 161, f. 164v–166).

[44] Pieter Biesboer, "Topographical Identifications for a Number of "Haerlempjes" by Jacob van Ruisdael," in *Shop Talk: Studies in Honor of Seymour Slive Presented on his Seventy-Fifth Birthday*, ed. Cynthia P. Schneider et al. (Cambridge, Mass, 1995), p. 36 and figs. 1, 5 and 6.

[45] Hans Renes, *Historische landschapselementen: Een lijst met definities en literatuur* (Wageningen, 1992), pp. 26–28; and *Les bocages: Historie, ecologie, economie* (Rennes, 1976).

very limited local fencing. Such a situation must have existed in the period of the hunter-gather cultures in prehistoric Europe. Later, farms, including possibly small gardens, were fenced off against all sorts of animals, but mainly cattle.[46] At the other end of the spectrum is the situation, where humans and a certain selection of (domestic, parasite) animals live together, whereas a number of specific (large, dangerous, amusing, exploited or rare) animals live in enclosed, fenced-off spaces. Such enclosures can vary from nature reserves to zoos to farms, but all are manipulated by humans, in more or less intensive ways. This is the situation in the present time.

The rabbit fences in the dunes of Holland in the period from 1500 to 1700 are situated somewhere in the middle of the spectrum: rabbits enclosed in relatively large areas, where the ecosystem was mildly manipulated by human agency. Yet animal fencing functioned on a different scale than before. No longer were individual houses or fields fenced off, but entire villages including all their agricultural land. In the dunes this new attitude to animal fencing was related to economic change and the new uses of the plains. Humans took full control over the beach plains and the leveled Old Dunes, and encroached upon the Younger Dunes, where the rabbits lived. These reflections lead to questions about continuity and change regarding the specific technology of rabbit fences. How did rabbit fences relate to other pre-existing fences? How did rabbits' natural behaviour affect the construction of fences and how did contemporary observers perceive that? How did the morphology of the land and its use influence the fences?

Rabbit fences in the Dutch dunes came in a large variety: hedges, earthen walls, fences made of wooden boards or wicker work, and ditches filled with (drainage) water. Several combinations existed, such as earthen walls topped with hedges or wooden fences, or ditches reinforced and lined by wooden fences. The vocabulary associated with these methods of rabbit control included rabbit fences (*konijnenheiningen*) and rabbit ditches (*konijnensloten*), but often more general words were used such as dune fences (*duinheiningen*) or territorial fences

---

[46] Survey books of archeology and reconstructions in open air museums and archeological parks such as the Archeon in Alphen aan den Rijn suggest this development, but I have not found any specific archeological literature on the topic. An example of a kraal (circular animal fence) dating from the period 3400–285 BC was found in the Netherlands near Anloo. See Theo Spek, *Het Drentse esdorpenlandschap: Een historisch-geografische studie: de belangrijkste resultaten*, 2 vols. (Utrecht, 2004), vol. 1, p. 127, n. 53.

(*banheiningen*). In Holland, rabbit fencing consisted of both dry and wet elements. This is indicated by an ordinance on the leasing out of rabbit rights issued in 1564 by the Holland Chamber of Accounts of the Habsburg government, which stipulated that landowners were allowed to build fences on their own lands, be it ditches, walls or other fences, under the condition that they had warned the dune lessee.[47]

In official documents only the word 'fence' might be used, but apparently that designated a more extensive construction, implying all sorts of means, including ditches. For instance, north of the IJ, the village of Bakkum was granted permission to build a rabbit fence in 1618, but in other documents about this fence the word 'rabbit ditch' appears. In 1777 when the construction was renovated, it turned out to consist mainly of ditches lined with fences.[48] The same applies to the nearby village of Castricum in 1659. Here ditches were part of the fencing system, as we may more aptly call it.[49] Similar combined fencing systems were used in Holland south of the IJ. In 1661 the dunes of the Lord of Wimmenum near The Hague were fenced, on one side by wooden boards and on the other side by a wide ditch. Combined means of fencing also occurred on the dunes of the Lord of Brederode near the city of Haarlem.[50]

The builders of dry rabbit fences had at least two pre-existing design examples at their disposal: cattle fences and game fences. Fences erected to keep cattle out of agricultural fields were often hedges made of thorny bushes. Such hedges had a long history.[51] The famous English landscape historian Oliver Rackham traced non-woodland trees and the word 'hedge' in English charters, the first dating from 980 A.D. Archaeological evidence about hedges in Great Britain seems to date

---

47 "Afheynige 't zy mit sloeten, dycken of andere heynigen … die by den eygenaers van den landen omtrent die voers. duynen leggende op haer selfs landt gedaen souden mogen worden dat midts den pachter gewaerschouwet hebbende." Ordonnance of 1564, The Hague, National Archive (NA), Grafelijkheidsrekenkamer Rekeningen [hereafter GRREK], inv. no. 399, f. 57 v.

48 By-law on the rabbit fence in Bakkum, Streekarchief Alkmaar, Gemeentebestuur Alkmaar, Gemeentebestuur Castricum, Oud-Archief, inv. no. 55, art. 2.

49 Approval of the Estates of Holland of the resolution of sheriff and aldermen of Castricum of 28 November 1659: Streekarchief Alkmaar, Gemeentebestuur Alkmaar, Gemeentebestuur Castricum, Oud-Archief, inv. no. L 27, arts. 1 en 3. Compare Jelles, *Geschiedenis*, p. 45.

50 Merula, *Placaten, p.* 188 (Wimmenum); and NA, GRRREG, inv. 723g, ff. 2 and 3v (Brederode dunes).

51 I thank J. Renes, Utrecht, for supplying me with some of the specialised literature on hedges.

back to at least 2000 B.C.[52] Among Dutch archaeologists a debate exists about the earliest evidence for hedges in the Netherlands. For a long time a hypothesis on the basis of pollen analysis posited that hedges date from the period that Neolithic humans penetrated the woods. A recent handbook survey on Dutch prehistory maintains that prehistoric field systems, formerly called *Celtic fields*, were lined with hedges.[53] A recent extensive in-depth interdisciplinary soil analysis by Theo Spek has found no trace of such early hedges. The first type of parcel demarcation seems to have been (wicker) wood fences.[54]

From the Roman period a reference to local practices in the Netherlands exists about making (temporary) hedges as war tools, so one can assume that the same techniques were used to make permanent, living hedges for more peaceful purposes. The Roman description is very similar to others based on contemporary evidence.[55] But the existence of a technique does not indicate conclusively that it was widely applied. The first occasional hedges appear in written sources such as farmer contracts dating from the 1300s. In the fifteenth century the number of references to maintenance obligations increased significantly. From the sixteenth century onwards, instructions about the construction of hedges were included in garden treatises. Sixteenth-century paintings and drawings like those by Pieter Breughel the Older are among the first pictorial sources showing living hedges. In the seventeenth and eighteenth centuries hedges were often depicted on maps. In particular in the river landscape of the Meuse hedges consisting of

---

[52] Oliver Rackham, *The Illustrated History of the Countryside* (London, 2003), p. 96. For the Netherlands no systematic investigation of medieval texts has been carried out concerning hedges. See *Inheemse bomen en struiken in Nederland en Vlaanderen: herkenning, verspreiding, geschiedenis en gebruik*, ed. Bert Maes (Amsterdam, 2006), p. 55.

[53] Willy Groenman-van Waateringe, "Hecken im westeuropäischen Frühneolithikum," *Berichten Rijksdienst voor het Oudheidkundig Bodemonderzoek* 20–1 (1970–1971), 295–9; and *Nederland in de prehistorie*, ed. Leendert P. Louwe Kooijmans et al. (Amsterdam, 2005), pp. 508, 547–9, 698, 707 and 709, with pl. 37. I thank C. C. Bakels, Leiden, for these references.

[54] Spek, *Het Drentse esdorpenlandschap*, pp. 702–4.

[55] Roman description: Caius Julius Caesar, *The Gallic War*, trans. J. J. Edwards, The Loeb Classical Library 72 (Cambridge, MA, 1917), par. 2.17.4: 'They cut into young saplings and bent them over, and thus by the thick horizontal growth of boughs, and by intertwining with them brambles and thorns, they contrived that these wall-like hedges should serve them as fortifications which not only could not be penetrated, but not even seen through." Contemporary descriptions: Gerry Barnes and Tom Williamson, *Hedgerow History: Ecology, History and Landscape Character* (Bollington, 2006), p. 2; and Paul Burm and Adriaan Haartsen, *Boerenland als natuur: Verhalen over historisch beheer van kleine landschapselementen* (Utrecht, 2003), pp. 62 and 138.

hawthorn (*meidoorn*) existed.[56] According to Spek, as long as fields were being extended by incorporating new, reclaimed parcels, wooden fences that were easy to remove and rebuild, played an important role. At the end of the Middle Ages, however, once the wilderness was mostly reclaimed and parcel boundaries became consolidated, it became worthwhile investing in more solid and permanent demarcations such as earthen walls with living hedges.[57] By the sixteenth century, in the western, most commercialized Netherlands, the maintenance of such thorny hedges seems to have become a profession. In 1564 the count of Holland leased out the maintenance of the hedges of his orchard and fields in The Hague to Jan and Aernt Cornelisz for seven years.[58]

Hedges were sufficient to keep cattle, horses and sheep off the fields, as they still are today in the eastern, high-lying sandy areas of the Netherlands. Keeping game within the borders of the game warren was much more capital-intensive, however, and elaborate constructions were needed. In an agreement about a deer fence made by the sheriff and aldermen of the village of Heemskerk dating from 1553 one finds the following information.[59] The foundation consisted of an earthen wall of 1.2 metres high, reinforced with (clay) sod or wood (the width is not given). On top of the wall a heavy wooden fence was built, without further specifications. This construction had passages with strong gates. This barrier probably had a total height of at least 2 metres. The height allows us to estimate the magnitude. In order to build a heavy fence on a earthen wall, the top must be at least half a metre wide. Estimating a slope of 45 degrees, the foundation must have easily been some 2.5-3 metres wide. Such a high and broad fencing construction would have been a prohibitive hindrance for any jumping animal.[60]

---

[56] Maes, *Inheemse bomen*, pp. 55–6; and J. Renes, *Landschappen van Maas en Peel: een toegepast historisch-geografisch onderzoek in het streekplangebied Noord- en Midden-Limburg*, Maaslandse monografieën 9 (Leeuwarden, 1999), p. 180.

[57] Spek, *Het Drentse esdorpenlandschap*, pp. 702–4.

[58] NA, GRRR, inv. 399, f. 91. As I once heard in an radio interview with a hedge specialist, knowledge is required about which species of plants are useful. The work consists partly of plaiting the branches and partly of planting new plants in the holes. In modern Dutch, such a person is aptly called a hedge-plaiter (*heggenvlechter*). In the seventeenth century, the word was thorn-knitter (*doornbreider*). See Alphons M. Hulkenberg, "Het kasboek van Jan van Matenesse, heer van Dever, 1587," *Jaarboekje voor geschiedenis en oudheidkunde van Leiden en omstreken* (1962), 52.

[59] Jelles, *Geschiedenis*, p. 36.

[60] In the sixteenth century in the sandy and hilly Veluweregion near the town of Wageningen in the central Netherlands, a comparable game fence existed that was more than 12 km long. It consisted of a dry moat and a wall topped with a wooden

Rabbit fences had to meet yet other requirements. Whereas most game jumps, rabbits also dig tunnels, so rabbit fences went deep into the ground. A rare description of a rabbit fence survives from 1668, when the above mentioned deer fence was remodeled. The inhabitants of the village of Heemskerk asked permission for a rabbit fence stretching along their municipal border with the dunes.[61] 0.3 metres of the fence was dug into the ground; 1.6 metres was above the ground. Costly wood was used. Since there was not much forest in the Netherlands, most wood was imported from Germany or even Scandinavia by the sixteenth century. So it was a wise decision of the inhabitants of Heemskerk to use recycled oak wood for the buried part of the fence, taken from demolished old ships, as the technical specification indicates. The upper part of the fence, however, had to consist of good (new) tightly fitting boards.[62]

The different requirements of fences related specifically to their function. Cattle fences had to be prickly and awkward to pass through, but could be made from local, living vegetative materials. In the Younger Dunes common sea-buckthorn (*Hippophae rhamnoides*, Dutch: *duindoorn*) must have been popular, as rules against the removal of this plant and others were issued by the forester of the Count of Holland.[63] Such hedges grew by themselves and only needed regular input of labor. Yet, building and maintaining a game fence was a much more costly affair. It had to be both high and wide and a large share of the material was (imported) wood. The rabbit fence, in contrast, was not wide, but high and it was inserted deeply into the ground so it needed a lot of labor input for both construction and digging. The total height of the wood structure (1.9 metres) must have meant an enormous capital input, just like the game fence. So rabbit fences were very similar to game fences in terms of capital-input, but differed in construction.

---

fence, an obstruction that totalled 2.8 m high. See F. W. J. Scholten, "Wildgraven, wildwallen en wildvreden op de Veluwezoom," *Historisch-Geografisch Tijdschrift* 16 (1998), 56–9; and Luuk Keunen and Hans Renes, "*... den wiltgraeff ofte wech van Wageingen naer Ede...*" *Resten van de wildwal aan de oostzijde van de Eng van Wageningen* (Wageningen and Utrecht, 2005).

[61] Jelles, *Geschiedenis*, p. 46.

[62] Studies on (still existing) boundaries of rabbit warrens show similar and other rabbit-proof features such as fences buried deep into the ground up to 0.5 m and steep walls lined at the foundation with stone to prevent burrowing. See Tom Williamson, *The Archaeology of Rabbit Warrens* (Princes Risborough, 2006), pp. 45–6.

[63] Merula, *Placaten*, p. 68 (reformation of 1524 of the *Placaet Generael opt stuck vande houtvesterie*, the so-called Statute of the Forestry, of 1517).

Unfortunately, a game fence was not sufficient to prevent rabbits from trespassing the warren borders.

In the dunes construction aimed at keeping the rabbits in the wilderness included ditches as well as fences. Ditches filled with drainage water could serve as a rabbit fence, but only under certain conditions. In particular the slope of the sides was crucial as a sixteenth-century series of conflicts about ditches in rabbit country reveals. The file includes the testimonies of eye-witnesses.[64] The villages involved were situated south of the IJ: Heemstede, Hillegom, Lisse, and Noordwijk en Noordwijkerhout. The imperial ordinance of 19 February 1550 ordained that the sides of the ditches between the dunes and the fields had to be slanted, not steep.[65] The reason was that the rabbits grazing along the ditches needed to be able to return to the dune warrens. The implication was that when the sides of the ditches were too steep, the rabbits would fall into the water and drown. The villagers defended themselves by maintaining that they kept the sides of the ditches steep on purpose, in order to prevent the rabbits from trespassing onto their fields.

Another argument is very remarkable and deserves consideration because it pertains to the perceived behaviour of rabbits and how the fences functioned, while at the same time providing great insight into contemporary nature observation. The villagers of Lisse claimed that rabbits did not fall into the water "because they do not drink and do not like water." The latter argument may seem somewhat suspicious, because in 1567 the inhabitants of the same village argued that mature rabbits did not drown, since they had often observed how they jumped into the ditchwater and jumped out again to walk back into the wilderness. But this apparent inconsistency can be understood by the fact that the villagers later specified that in the spring *young, inexperienced* rabbits sometimes drowned.[66] The testimonies about swimming rabbits are remarkable, since the renowned Dutch biologist and rabbit specialist Marijke Drees has stated that she has never come across any references in literature to swimming rabbits. She has spent much time observing rabbits in the dunes and has never seen a swimming rabbit herself.

---

64 The entire court file is published in *Inventaris en beschrijving van de processtukken (dossiers) behorende tot de Beroepen uit Holland berustende in het archief van de Grote Raad van Mechelen*, ed. Andries, J. C. et al., 10 vols. (Amsterdam, 1964–74), Inventaris I, dossier 66bis. Parts of the file have also been published in Merula, *Placaten*, pp. 186–94.

65 Merula, *Placaten*, p. 186.

66 Hulkenberg, *Keukenhof*, p. 6.

Even rabbits hunted by stoats do not flee by taking a route through water. Healthy rabbits usually take well-known land routes.[67]

Some fences combined water and dug-in fences, as described in the aforementioned renovation of the fence of Bakkum in 1777. On the side of the fields, the fence stood in a ditch. Presumably, one should take into account that fences functioned differently according to the season. There are reports that in the dry season the ditches were filled in by sand as a consequence of sand drift. So, in the summer, ditches caused no obstacle to rabbits. The same applies to severe winters, when water-filled ditches froze. We must conclude that rabbit fences could be very complex constructions combining dry and wet components so as to accommodate for different seasons and local varieties in the landscape.

The conflicts about the ditch sides made it to the Great Council of Malines, the highest Law Court and Court of Appeal of the Habsburg Netherlands. Following the results of local research by the Great Council, Emperor Charles V sought a compromise.[68] In the ordinance of 16 February 1553, he adapted the technical specifications of the sides of the ditches.[69] He stipulated three standard slopes, depending on the local height of the dunes. The owners of the lands were responsible for maintenance of these norms, since the ditches were dug at their cost.

Conflicts about the slope of the sides of the ditches led to new legislation again at the end of the sixteenth century (after the successful Dutch Revolt against the Habsburg government). In the ordinance of 10 September 1597, the new government, the Estates of Holland, came up with a new solution to the problem of the steep ditch sides.[70] It ordered the landowners to construct, at regular distances of four to five metres, ramps with a width of 0.6 metre. We can presumably imagine these ramps built into the steep sides of the ditches as roads built into mountain sides. This solution proved to be more enduring. The only new item that gave rise to discussion thereafter was who had to finance

[67] Email correspondence from M. Drees (9–10–2007).

[68] For the verdict, see *Chronologische lijsten van de geëxtendeerde sententiën en procesbundels (dossiers) berustende in het archief van de Gote Raad van Mechelen*, ed. Jacobus Th. de Smidt and E. I. Strubbe, 6 vols. (Brussels and Utrecht, 1966–1988), vol. 5, no. 1813; and *Haarlemers voor de Grote Raad van Mechelen 1458–1578: Gebaseerd op Algemeen Rijksarchief Brussel fonds Grote Raad en Stadsarchief van Haarlem*, ed. Jacobus Th. de Smidt (Haarlem, 1999), p. 98, no. 169.

[69] Merula, *Placaten*, p. 188.

[70] Ibid., p. 192.

the maintenance of this micro-environmental infrastructure: the landowners or the land users. A 1642 court ruling decided to keep the landowners responsible for this.[71]

When the canals were broad enough, no fear for swimming rabbits existed. From the agrotown of Beverwijk a very detailed map survives, showing all of its vegetable gardens. The gardens extend to the border of the village, which consisted of a broad waterway called *beek*. On the other side of the water the wild dunes are visible. In his commentary to the map, the board secretary of the community of Beverwijk wrote that the village was decorated by beautiful dunes enriched by rabbits and very suitable for walks. Presumably, the gardeners of Beverwijk felt protected from the rabbits by their broad waterway.

## *Conclusion*

This study on rabbit fences in the coastal dunes of the Netherlands shows how changes in the landscape and in land use affected the relationship between humans and animals. The dunes were a peculiar landscape of sand hills, valleys and peat covered plains. In the Middle Ages rabbits were introduced and stimulated to thrive in designated warrens, for the purpose of meat and fur production. The feral rabbits cohabitated peacefully with cattle belonging to peasants in nearby villages. In this almost Arcadian situation no rabbit fences existed. From 1400 onwards, however, parts of the dunes were leveled for peat and sand mining, and both old and new plains were drained and reclaimed for agriculture and industry, such as intensive horticulture and linen bleaching. The new entrepreneurs had a strong interest in keeping rabbits off their premises and they had sufficient capital to invest in fencing. In the early modern period the old practice of offgrazing beyond the rabbit warrens gave way to the building of elaborate rabbit fences, which came in all forms and shapes. Earthen walls and fences made of wooden boards and wicker work appear to have existed all along the dunes. Often they were combined in one construction with ditches serving both as drainage canals and as parcel demarcations. The watery element was particularly prevalent in Holland south of the IJ, where extensively drained peaty lands formed an important feature of the

[71] Act of 13 July 1642 (Streekarchief Kennemerland, Archief heerlijkheid Heemstede, inv. no. 283).

dune landscape. The rabbit fences were more costly in terms of effort, labour, and capital than the pre-existing cattle and game fences. Humans were very well aware of how rabbits operated in the landscape and dealt with the micro-environmental infrastructures, depending on place, age and season, and they adapted their choice of fencing system accordingly. Even for swimming rabbits it became very difficult to cross the borders of their warrens.

## *Bibliography*

Augustyn, Beatrijs. *Zeespiegelrijzing, transgressiefasen en stormvloeden in maritiem Vlaanderen tot het einde van de XVIde eeuw: Een landschappelijke, ecologische en klimatologische studie in historisch perspectief*, 2 vols. (Brussels, 1992).

Bakker, Boudewijn Bakker. "De zichtbare stad 1578–1813," in *Geschiedenis van Amsterdam: Centrum van de Wereld, 1578–1650*, ed. Maarten Prak and Willem Frijhoff (Amsterdam, 2005), pp. 17–48.

Bakker, Elisabeth S. *Herbivores as Mediators of their Environment: The Impact of Large and Small Species on Vegetation Dynamics* (Wageningen, 2003).

Barnes, Gerry and Tom Williamson. *Hedgerow History: Ecology, History and Landscape Character* (Bollington, 2006).

Beenakker, Jan J. J. M. *Lisse op de grens van droog en nat. De bewoningsgeschiedenis en landschapsontwikkeling van een dorp in de bloembollenstreek tot omstreeks 1900* (Lisse, 1993).

Bieleman, Jan. *Geschiedenis van de landbouw in Nederland 1500–1950: veranderingen en verscheidenheid* (Meppel, 1992).

Biesboer, Pieter. "Topographical Identifications for a Number of "Haerlempjes" by Jacob van Ruisdael," in *Shop Talk: Studies in Honor of Seymour Slive Presented on His Seventy-Fifth Birthday*, ed. Cynthia P. Schneider et al. (Cambridge MA, 1995), pp. 36–39, with plates 286–88.

Blok, D. P. et al. eds. *Algemene Geschiedenis der Nederlanden* (Haarlem, 1977–83).

*Les bocages: Historie, ecologie, economie* (Rennes, 1976).

Boer, Dick E. H. de. *Graaf en grafiek. Sociale en economische ontwikkelingen in het middeleeuwse 'Noordholland' tussen 1345 en 1415* (Leiden, 1976).

Borger, Guus J. "Draining-Digging-Dredging: The Creation of a New Landscape in the Peat Areas of the Low Countries," in *Fens and Bogs in the Netherlands: Vegetation, History, Nutrient Dynamics and Conservation*, ed. Joseph T. A. Verhoeven (Dordrecht, 1992), pp. 131–71.

Burm, Paul and Adriaan Haartsen. *Boerenland als natuur: verhalen over historisch beheer van kleine landschapselementen* (Utrecht, 2003).

Caesar, Caius Julius. *The Gallic War*, trans. J. J. Edwards, The Loeb Classical Library 72 (Cambridge, MA, 1917).

*Chronologische lijsten van de geëxtendeerde sententiën en procesbundels (dossiers) berustende in het archief van de Gote Raad van Mechelen*, ed. Jacobus Th. de Smidt and E. I. Strubbe, 6 vols. (Brussels and Utrecht, 1966–88).

Cornelisse, Charles. *Energiemarkten en energiehandel in Holland in de late middeleeuwen* (Hilversum, 2008).

Dam, Petra J. E. M. van. "Sinking Peat Bogs: Environmental Change in Holland, 1350–1550," *Environmental History* 5 (2000), 32–45.

——. "De rol van de warande: Geschiedenis en inburgering van het konijn," *Jaarboek Ecologische Geschiedenis 2000* 5 (2002), 59–84.

——. "Ein Neubürger in Nordeuropa: Menschliche und natürliche Einflüsse auf die Assimilierung des Kaninchens in den Niederländischen Dünen 1300–1700," in *Beiträge zum Göttinger Umwelthistorischen Kolloquium 2004–2006*, ed. Bernd Herrmann (Göttingen, 2007), pp. 167–80.

Dekker, Jasja J. A. *Rabbits, Refuges and Resources: How Foraging of Herbivores is Affected by Living in Burrows* (Wageningen, 2007).

Doorn, Maarten van and Jaap Mennema. *De Haagse Beek: een (natuur)historische* verkenning (The Hague, 1992).

Fockema Andreae, Sybrandus J. *Het Hoogheemraadschap van Rijnland. Zijn recht en zijn bestuur van den vroegsten tijd tot 1857* (Leiden, 1934; repr. Alpen aan den Rijn, 1982).

*Generale Privilegiën van Kennemerland, etc.* (The Hague, 1665).

*Geschiedenis van Amsterdam*, ed. Maarten Prak and Willem Frijhoff, 4 vols. (Amsterdam, 2005).

Groenman-van Waateringe, Willy. "Hecken im westeuropäischen Frühneolithikum," *Berichten Rijksdienst voor het Oudheidkundig Bodemonderzoek* 20–21 (1970–71), 295–99.

Groesbeek, Johan W. *Middeleeuwse kastelen van Noord-Holland: hun bewoners en bewogen geschiedenis* (Rijswijk, 1981).

*Haarlemmers voor de Grote Raad van Mechelen 1458–1578: Gebaseerd op Algemeen Rijksarchief Brussel fonds Grote Raad en Stadsarchief van Haarlem*, ed. Jacobus Th. de Smidt (Haarlem, 1999).

Harten, J. D. H. "Stedelijke invloeden op het Hollandse landschap in de 16e, 17e en 18e eeuw," *Holland: regionaal-historisch tijdschrift* 10 (1978), 114–34.

Hoekstra, Pier. *Bloemendaal, proever ener streekgeschiedenis* (Wormerveer, 1947).

Hulkenberg, Alphons M. "Het kasboek van Jan van Matenesse, heer van Dever, 1587," *Jaarboekje voor geschiedenis en oudheidkunde van Leiden en omstreken* (1962), 50–63.

——. *Keukenhof*, Hollandse studiën 7 (Dordrecht, 1975).

*Informacie up den staet, facultueyt ende gelegentheyt van de steden ende dorpen van Hollant ende Vrieslant om daernae te reguleren de nyeuwe schiltaele gedaen in jare MDXIV*, ed. Robert Fruin (Leiden, 1866).

*Inheemse bomen en struiken in Nederland en Vlaanderen: herkenning, verspreiding, geschiedenis en gebruik*, ed. Bert Maes (Amsterdam, 2006).

*Inventaris en beschrijving van de processtukken (dossiers) behorende tot de Beroepen uit Holland berustende in het archief van de Grote Raad van Mechelen*, ed. J. C. Andries et al., 10 vols. (Amsterdam, 1964–74).

Jelgersma, Saskia et al., "The Coastal Dunes of the Netherlands: Geology, Vegetational History and Archaeology." *Mededelingen Rijksgeologische Dienst*, n.s. 21 (1970), 93–167.

Jelles, J. G. G. *Geschiedenis van beheer en gebruik van het Noordhollands Duinreservaat* (Arnhem, 1968).

Jong, Erik de. "For Profit and Ornament: The Function and Meaning of the Dutch Garden Art in the Period of William and Mary 1650–1702," in *The Dutch Garden in the Seventeenth Century*, ed. John D. Hunt (Washington D.C., 1990), pp. 13–48.

Keunen, Luuk and Hans Renes, "*... den wiltgraeff ofte wech van Wageingen naer Ede...*" *Resten van de wildwal aan de oostzijde van de Eng van Wageningen* (Wageningen and Utrecht, 2005).

Koningsveld et al., M. van. "Living with Sea-Level Rise and Climate Change: A Case Study of the Netherlands," *Journal of Coastal Research* 24 (2008), 395–407.

Kops, Jan. *Staat van Holland's duinen en ontwerp tot vrugtbaarmaking van dezelve, opgegeven in het rapport der comissie over het onderzoek der duinen van Holland. Tegenwoordige staat der duinen (van het voormalige Holland)*, vol. 1 of 2 vols., 2nd ed. (Utrecht, 1818).

Kreike, Emmanuel "Architects of Nature: Environmental Infrastructure and the Myth of Natural Resources Management," Paper presented at the Fourth Conference of the European Society of Environmental History, Amsterdam, The Netherlands, June 5–9, 2007.

Ligtendag, Willem A. "De kustlijn in kaart," *Geografisch tijdschrift* 5 (1990), 380–6.

Lucassen, Jan. *Naar de kusten van de Noordzee: Trekarbeid in Europees perspektief, 1600–1900* (Gouda, 1984).

*Man-Made Lowlands: History of Water Management and Land Reclamation in the Netherlands*, ed. Gerardus P. van de Ven (Utrecht, 1993).

McNeill, John. *Something New Under the Sun: An Environmental History of the Twentieth Century* (London, 2000).

Merula, Paulus G. F. P. N. *Placaten ende ordonnancien op 't stuck van de wildernissen* (The Hague, 1605).

Monté Ver Loren, Johan Ph. de, and Johannes E. Spruit, *Hoofdlijnen uit de ontwikkeling der rechterlijke organisatie in de Noordelijke Nederlanden tot de Bataafse omwenteling* (Deventer, 1982).

*Nederland in de prehistorie*, ed. Leendert P. Louwe Kooijmans et al. (Amsterdam, 2005).

*Prins Maurits' kaart van Rijnland en omliggend gebied door Floris Balthasar en zijn zoon Balthasar Florisz van Berckenrode in 1614 getekend*, ed. Kees Zandvliet (Alphen, 1989).

Rackham, Oliver. *The Illustrated History of the Countryside* (London, 2003).

Radkau, Joachim. *Natur und Macht: Eine Weltgeschichte der Umwelt* (Munich, 2002); English translation: *Nature and Power: A Global History of the Environment* (Cambridge, 2008).

*Rechtsbronnen der stad Haarlem*, ed. Johan Huizinga (The Hague, 1911).

Regtdoorzee Greup-Roldanus, S. C. *Geschiedenis der Haarlemmer bleekerijen* (The Hague, 1936).

Renes, Hans. *Historische landschapselementen: een lijst met definities en literatuur* (Wageningen, 1992).

Renes, Johannes. *Landschappen van Maas en Peel: een toegepast historisch-geografisch onderzoek in het streekplangebied Noord- en Midden-Limburg*, Maaslandse monografieën 9 (Leeuwarden, 1999).

Rentenaar, Rob. "De Nederlandse duinen in de Middeleeuwse bronnen tot omstreeks 1300," *Geologisch Tijdschrift* 11 (1977), 361–76.

——. "De vroegste geschiedenis van het konijn in Holland en Zeeland," *Holland: regionaal-historisch tijdschrift* 10 (1978), 2–16.

Sangers, W. J. *De ontwikkeling van de Nederlandse tuinbouw* (Zwolle, 1952).

Scholten, F. W. J. "Wildgraven, wildwallen en wildvreden op de Veluwezoom," *Historisch-Geografisch Tijdschrift* 16 (1998), 56–9.

Schoorl, Henk. *De Convexe Kustboog. Texel-Vlieland-Terschelling 1: Het westelijke Waddengebied en het eiland Texel tot circa 1550* (Schoorl, 1999).

——. *Kust en kaart. Artikelen over het kaartbeeld van het Noordhollandse kustgebied* (Schoorl, 1990).

Sloof, Jan H. M. "Rijnland en het Spaarne in de late middeleeuwen," in *De loop van het Spaarne: de geschiedenis van een rivier*, ed. Bert C. Sliggers (Haarlem, 1987), pp. 17–35.

Spek, Theo. *Het Drentse esdorpenlandschap: een historisch-geografische studie: de belangrijkste resultaten* (Utrecht, 2004).

*The State of American History*, ed. Herbert J. Bass (Chicago, 1970).

Thurkow, Albert J. "Het agrarisch landgebruik in de Hollandse duinen in historisch perspectief," *Duin: contactblad van Stichting Duinbehoud en Waterwinning* 3 (1986), 75–8.

Tielhof, Milja van, and Petra J. E. M. van Dam. *Waterstaat in stedenland. De geschiedenis van het hoogheemraadschap van Rijnland voor 1857* (Utrecht, 2006).
*De tuin van Holland. Geschiedenis van tuinbouw en groenteveiling in Leiden*, ed. Johannes G. Endhoven et al. (Utrecht, 1992).
Vries, Jan de. *The Dutch Rural Economy in the Golden Age, 1500–1700* (London, 1974).
Wallace, Alfred. *The Wonderful Century, Its Successes and Its Failures* (London and New York, 1898).
Williamson, Tom. *The Archaeology of Rabbit Warrens* (Princes Risborough, 2006).
Winter, Johanna M. van. "Green Salads: An Innovation in the Diet of the Renaissance Period," in *Spices and Comfits: Collected papers on Medieval Food*, ed. Johanna M. van Winter (Totnes, 2007), pp. 369–81.
Zagwijn, Waldo H. "The Formation of the Younger Dunes on the West Coast of the Netherlands (AD 1000–1600)," *Geologie en Mijnbouw* 63 (1984), 259–68.
Zeiler, Frits D. *Nollen, krochten, blinken: duintoponiemen tussen Wijk aan Zee en Camperduin* (Castricum, 1995).
Zeischka, Siger. *Minerva in de polder: waterstaat en techniek in het hoogheemraadschapvan Rijnland (1500–1856)*, Ph.D. thesis, Vrije Universiteit, (Hilversum, 2008).
Zuurbier, S. P. A. "De heerlijkheid Bakkum en zijn ambachtsheerlijkheid," *Oud-Castricum: Jaarboekje* 3 (1980), 3–18.

# THE ART OF MAKING THE EARTH FRUITFUL: MEDIEVAL AND EARLY MODERN IMPROVEMENTS OF SOIL FERTILITY

Verena Winiwarter

Many arrivals make us live: the tree becoming
Green, a bird tipping the topmost bough,
A seed pushing itself beyond itself,
The mole making its way through darkest ground,
The worm, intrepid scholar of the soil—…
Theodore Roethke (1908–1963), *The Manifestation*

## *How to Sustain Agriculture: Gathering Experience with the Soil*

A little known agricultural manual from the ninth century AD, written in the then flourishing Abbasid Caliphate in Baghdad, is not untypical for two millennia of systematic gathering of agricultural knowledge and its interpretation in the light of the theory of the times. Ibn Wahsiyya's compilation "The Nabatean Agriculture" comments on experiments conducted with a fertilizer made from the ash of a special kind of olive and ammonia.[1] The yellowish salt that results from mixing the two components was to be added to the water used for plants. According to Ibn Wahsiyya, the difference the fertilizing liquid makes can easily be seen by comparing two plants, one treated with such water, the other one without. The treated one will grow faster and reach a greater height than the other. Remarkably, this ninth-century tract refers to what can only be called a straightforward parallel experiment. Such methodical rigor is not unusual in agricultural manuals. The noted effects of interventions were explained in theoretical terms radically different from those applied today, but we should bear in mind that the empirical evidence was gathered systematically.

---

[1] All references are quoted from Ernst Bergdolt, "Beiträge zur Geschichte der Botanik im Orient. I. Ibn Wahhschija: Die Kultur des Veilchens (*Viola odorata* L.) und die Bedingungen des Blühens in der Ruhezeit," *Berichte der Deutschen Botanischen Gesellschaft* 52 (1934), 87–94; reprinted in *Natural Sciences in Islam*, ed. Fuat Sezgin (Frankfurt/Main, 2001), pp. 249–64.

Much has been written about the history of agricultural implements, plant cultivation, animal husbandry and the organization of cultivation, its development from a two- to a three-field system, and then on to systems of crop rotation. Much less has been written about the central problem incurred when colonizing the earth, the problem Ibn Wahsiyya is talking about: to render and keep the earth fertile. Soils have been curiously neglected in the history of agriculture and in environmental history alike. Edward John Russell's *Soil Conditions and Plant Growth*, the great classic of its genre, was first published in 1912 and has since gone through a dozen editions. It starts with a remarkably learned historical overview.[2] Of its thirty pages, eighteen are used to describe historical development prior to 1860. But the historical overview commences with "The Search for the Principle of Vegetation, 1630–1750" as the first full chapter, whereas all earlier history is summarized in one printed page. Russell writes the history of fertility theories from a viewpoint of perceived elucidation peering into darker ages and seeks the roots of contemporary understanding. His chapter is nevertheless still among the best existing overviews.

Recent contributions by Christian Feller and his co-workers, who probed more deeply into the history of fertility concepts of the nineteenth century are conceptually different. They were looking for quantitative fertility indicators to evaluate the sustainability of cropping and farming systems. Albrecht Daniel Thaer's famous volumes on the principles of rational agriculture were the basis of this investigation.[3] Another laudable historical study focused on Charles Darwin's work on the role of earthworms in the formation of vegetable mould.[4] But Feller and his colleagues work on the nineteenth century. This paper takes a longer look. As part of a larger project on the environmental history of soils, this study traces knowledge about ways to make the earth fruitful from the early medieval period to the eighteenth century.[5]

---

[2] *Russel's Soil Conditions and Plant Growth*, ed. Alan Wild, 11th ed. (Essex, 1988).

[3] Christian Feller et al., "The Principles of Rational Agriculture by Albrecht Daniel Thaer (1752–1828): An Approach to the Sustainability of Cropping Systems at the Beginning of the Nineteenth Century," *Journal of Plant Nutrition and Soil Science* 166 (2003), 687–98.

[4] Christian Feller et al., "Charles Darwin, Earthworms and the Natural Sciences: Various Lessons From the Past to Future," *Agriculture, Ecosystems and Environment* 99 (2003), 29–49.

[5] For other contributions to this larger attempt, see *Soils and Societies: Perspectives from Environmental History*, ed. J. R. McNeill and Verena Winiwarter (Strond, 2006); and J. R. McNeill and Verena Winiwarter, "Breaking the Sod: Humankind History and Soil," *Science* 304, no. 11 (2004), 1627–9.

Conceptual clarification of the notion of soil fertility is an important basis for the history of fertility.[6] The term "fertility" will be used hereafter as it is conceptualized in the sources presented, and no attempt at a unified fertility theory will be made; rather, the diversity is what is of interest.

Agriculture was central to pre-modern societies. Despite being considered the peasants' way of living, less esteemed than the noble pursuit of hunting, agricultural knowledge was part of the essential and ubiquitous body of knowledge of pre-modern societies, considered self-evident in a way that might perhaps be compared to the self-evident character of knowledge about personal computer programs today. Agricultural knowledge was (dis)qualified as experiential rather than experimental in the nineteenth century in the emerging sciences' quest for the establishment of scientific disciplines.[7] This has hampered due recognition of the existing advancements.

While the examples presented below stem from specialized works, the history of knowledge about soils should not be confined to what is commonly, and somewhat misleadingly, called "technical literature." Knowledge about and appreciation of soils was part of the culture of Europe since the advent of writing. A reading of the parables of the Bible and the exegetical literature, in which agricultural analogies abound, should suffice to show how deeply ancient culture was embedded in its agriculture.[8] It was common knowledge in pre-industrial times that everything – the wealth and prosperity of nations – depended on the fertility of the soil, a way of seeing the world which has since been disqualified as "environmental determinism," despite its insight into the dependence of humankind on ecosystem services. Soil fertility was a major concern of Mercantilist economists and the school of the Physiocrats in general. To give but one example, the well-known French philosopher of law and practicing landowner Charles de Secondat, Baron de Montesquieu (1689–1755), wrote the seminal "L'Esprit des lois," which was published in 1748 after the author had traveled widely

---

[6] See the review article by Nikola Patzel, Hans Sticher, and Douglas L. Karlen, "Soil Fertility: Phenomenon and Concept," *Journal of Plant Nutrition and Soil Science* 163 (2000), 129–42.

[7] Soraya de Chadarevian, "Laboratory Science Versus Country-House Experiments: The Controversy Between Julius Sachs and Charles Darwin," *The British Journal of the History of Science* 29 (1996), 17–41.

[8] Winfried E. H. Blum and Verena Winiwarter, "Souls and Soils: A Survey of Worldviews," in *Footprints in the Soil: People and Ideas in Soil History*, ed. Benno Warkentin (Amsterdam, 2006), pp. 107–22.

through Europe.[9] In Book XVIII. "Of Laws in the Relation They Bear to the Nature of the Soil," he commences with an historical analysis of the interaction between nature and society. He holds that the barrenness of the earth renders men industrious, sober, inured to hardship, courageous, and fit for war, while fertile soils result in less hardy peasants. Montesquieu is an excellent example of mid-eighteenth-century political thought, in which soil conditions played a central role. He sees a strong connection between the makeup of a society and its natural resources, and in that he is principally right. In the solar-based society he analyzed, soils were indeed of paramount importance.

Getting back to more specialized literature, Russell's distinction between a first period of systematic searches for soil fertility from the 1630s to ca. 1750 and the phlogistic period thereafter is useful, but much remains to be said, in particular concerning the learned practice of agriculture. Soil fertility theories did not exist outside of a social context. Agriculture was considered to be a practical field worth detailed, systematic study as opposed to inductive theorizing. In a review of a work on soil fertility published in 1791, the reviewer summarizes coldly that some errors in the work, which in general he considers useful, are due to the fact that the author is more a theoretical mind than a practitioner. The publication in which this review appeared, the *Allgemeine Deutsche Bibliothek*, was aimed at a broader public, and therefore such a judgment was directed at a larger audience.[10] Studying the history of soil knowledge thus allows us – the usual caveats with regard to literacy and the paucity of books notwithstanding – to recapture systematic, practical knowledge, knowledge which might be termed 'experiential', of the kind which has been identified as necessary for sustainable agriculture.[11] What exactly is this knowledge about? It is not about "nature" as such, but about the interaction between nature and human culture.

Nature has been colonized by humans for thousands of years by means of agriculture. The notion of colonizing interventions, developed in social ecology, can be used to describe the agricultural relationship

---

[9] Charles de Montesquieu, *De l'Esprit de loi* (Geneva, 1748). Originally published anonymously.

[10] Review by C. W. F. L. Frhr. von Poellniz, "Über die wahren Mittel zur Fruchtbarkeit, eine physikalisch-ökonomische Abhandlung," *Allgemeine deutsche Bibliothek* 102 (1791), 463. The original quote reads: "Einige Fehler in diesem an sich nützlichen Werke mögen wohl daher rühren, daß der Verf. mehr theoretischer als praktischer Oekonom ist."

[11] Lars Ohlander, Charlotte Lagerberg, and Ulla Gertsson, "Visions for Ecologically Sound Agricultural Systems," *Journal of Sustainable Agriculture* 14 (1999), 73–9.

between nature and culture and the environmental consequences of agriculture.[12] Such interventions are designed to use natural organisms or ecosystems to produce biomass deemed useful by humans. Colonizing, in contrast to producing artifacts, is achieved through continued use of natural processes, such as photosynthesis, steering them in the desired direction, for example, promoting grains by sowing, manuring and weeding its natural competitors, in contrast to burning clay to make a pot or cutting wood to make a table, where the natural system is no longer used in its productive capacity. Colonizing interventions can take place at the level of the single organism, as when humans select animals or plants for breeding. They can also take place at the ecosystem level in the fields, meadows, olive-groves and vegetable gardens of the agriculturalists. Colonizing has been successful over thousands of years in increasing the amount of biomass *available for humans* per area, but it comes at a price. One enters a spiral of risk as successful coping strategies invariably produce more, and other, new risks,[13] Moreover, colonizing means work, in terms of constant care of the cultivated, colonized parts of the earth. Such stewardship is a complicated task and needs to be based on experience, diligence and patience. The one who plants the olive tree will not necessarily be the one reaping the olives, meliorating the soil for a vineyard will be effective only after several years, and the prevention of erosion cannot be seen at all, since only the failure to prevent it becomes visible. Such stewardship was once a mainstay of collective wisdom – although it was certainly not translated into corresponding practice everywhere.

With renewed interest in such collective wisdom as part of the search for a sustainable future it has come to be called traditional ecological knowledge or TEK for short.[14] For the pursuit of a sustainable future, looking back into the knowledge base of pre-industrial, solar-based,

---

[12] Marina Fischer-Kowalski et al., *Gesellschaftlicher Stoffwechsel und Kolonisierung von Natur: Ein Versuch in Sozialer Ökologie* (Amsterdam, 1997), pp. 10–12.

[13] Rolf Peter Sieferle and Ulrich Müller-Herold, "Surplus and Survival: Risk, Ruin, and Luxury in the Evolution of Early Forms of Subsistence," *Advances in Human Ecology* 6 (1997), 201–20.

[14] It was Richard Hoffmann who first saw the value of applying the TEK concept to medieval knowledge of agriculture, for example, in presentations at the Social Science History Association conference in 2006 and the European Society for Environmental History conference in 2007. The concept is detailed in International Council for Science, "Science, Traditional Knowledge and Sustainable Development," ICSU Series on Science for Sustainable Development 4 (Paris, 2002).

and hence potentially more sustainable societies is deemed useful.[15] Without going into the details of a vast literature, sustainability can be understood as a way of dealing with natural systems that does not compromise the integrity of these, hence allowing for continued interaction over long periods. The recent interest in traditional ecological knowledge has mainly been connected with indigenous societies, somewhat arbitrarily excluding those who did write down their experiences.[16] But the historical record of European agricultural treatises contains an enormous amount of knowledge which should be considered as TEK.

It has been argued that a creative combination of scientific knowledge, technology, experience and common sense would be necessary to design agricultural systems that are ecologically sound.[17] Under current conditions in education and consulting, experience and common sense are hardly appreciated. To achieve sustainability, it might be worth trying to bring an appreciation of experience-based knowledge back into our education systems. One way to do so is to search for counter-narratives to the prevailing enlightenment and progress stories about the advent of scientific agriculture. The history of soil fertility management in pre-industrial agriculture is one such counter-narrative.

The agricultural literature, from which such stories can be taken, is vast, and one first needs to differentiate the form and intended audience of the works. Richard Hoffmann discerns isolated recipes, unstructured memoranda, brief, organized tracts and consciously integrated treatises as main categories for late medieval writings on fishing.[18] In antiquity, agricultural knowledge was offered in tracts such as Cato's and Varro's books, but also in well-organized treatises such as those of Columella or Palladius. In medieval times we find agricultural knowledge incorporated into encyclopedias, well-organized but rather impractical. The agricultural writings of early modern Europe take the form of tracts (such as Grosser and Tumbshirn, and perhaps also the Haushaltungsbüchlein of the Grünthaler family in Upper Austria,

---

[15] Verena Winiwarter, "Agrarökologische Geschichtsschreibung: Der Umwelthistorische Beitrag Zur Diskussion Um Nachhaltige Agrar-Entwicklung," *GAIA* 11 (2002), 104–12.

[16] See, for example, Akin Omotayo and M. W. Musa, "The Role of Indigenous Land Classification and Management Practices in Sustaining Land Use System in the Semi-Arid Zone of Nigeria," *Journal of Sustainable Agriculture* 14 (1999), 49–58.

[17] Ohlander, "Visions for Ecologically Sound Agricultural Systems," p. 73.

[18] Richard C. Hoffmann, *Fishers' Craft and Lettered Art: Tracts on Fishing from the End of the Middle Ages* (Toronto, 1997), p. 322.

unpublished in its time) or treatises, such as Coler's *Oeconomia*, and Florinus' or Helmhard von Hohberg's works and many others categorized as "Hausväterliteratur."

Agricultural treatises have either been organized by theme, by time, or in combination of both systems.[19] Soil knowledge was also communicated in books generally organized by time, such as calendars, where it did not fit easily into the composition. In Palladius' influential fourth-century book, soils are the theme of the second of 14 chapters, 12 of which are dedicated to a month each. Soils had to be treated in different contexts, such as agriculture proper, vineyards, orchards, gardens and ponds, and hence, posed a formidable organizational problem to authors.

Looking into classical, medieval and early modern treatises on agriculture, prescriptions that look ecologically sound from our current point of view are found next to recipes for magical potions, a fact that makes evaluation of the knowledge difficult. My judgment on this matter is impressionistic, based on years spent with the sources. My impression is that in most cases, magical recipes were given for ailments for which no other remedies were known, for pests, for diseases in humans and animals, and in connection with the vagaries of the weather. At the same time, study of these phenomena continued, and eventually some of the riddles were solved and more rational treatments ensued. Altogether, in the agricultural literature, the best available theoretical explanations were offered. From the Greek natural philosopher Empedokles up to the mid-nineteenth century – transmitted to later times mainly via Aristotle – humor theory, the theory of the four elements earth, water, fire and air, in connection with the four antagonistic qualities wet/dry and warm/cold, formed the theoretical framework of European and Arabic agricultural knowledge. From the late sixteenth century onwards this foundation was increasingly mingled with bits and pieces of alchemical theory. Our theories are different now, but what remains is an altogether astonishing collection of experiences, often gathered so systematically that these practices should be considered experiments. A few of the lesser known examples with regard to knowledge about soils, the prerequisite for agriculture *par excellence*, will illustrate this point.

---

[19] Verena Winiwarter, "Zwischen Himmel und Erde: Umwelthistorische Aspekte landwirtschaftlicher Kalender," in *Ideologisierte Zeit: Kalender und Zeitvorstellungen im Abendland von der Antike bis zur Neuzeit*, ed. Wolfgang Hameter, Meta Niederkorn-Bruck, and Martin Scheutz, Querschnitte 17 (Innsbruck, 2005), pp. 276–94.

### *A Journey from Byzantium to Hamburg: From the Geoponika to Eller's Treatise on the Fertility of the Soil*

This part of my overview presents a few stepping stones in the vast corpus of soil knowledge. It tries to achieve two things: to give a general impression of the types of intervention prescribed in the technical literature as means to make the earth fruitful; and to offer some insight into the development of theories of fertility. Both the starting and stopping point of this journey are somewhat arbitrary. Stepping stones on the journey are not to be mistaken for a line of direct interconnection, but represent nodules in a web of knowledge that remains largely uncharted. The three examples were chosen for diversity, which renders them unfit for any direct comparison. They were written under distinctly different natural conditions. Mediterranean agriculture is challenged by a lack of moisture, whereas temperatures are usually no problem and sunshine is abundant. Torrential rainfalls result in severe soil erosion. Desiccation and the formation of hard layers of impenetrable dried clays threaten soils. In most places of temperate Europe, rainfall and thus soil moisture are unproblematic, whereas agriculture is limited by low temperatures. The difference is mirrored in agricultural techniques such as lithic mulching for water conservation in the Mediterranean or the moldboard plough in temperate Europe, which allows the cultivation of heavy, wet soils by turning the sods to dry them.

For the Mediterranean case, knowledge about soils and the means to sustain their fertility that were collected by ancient Roman writers has been discussed in some detail.[20] Less has been published about the south-eastern European tradition, which was written in Greek as opposed to Latin. The *Geoponika*, a Byzantine collection of otherwise lost agricultural writings, were, according to current scholarly opinion, compiled in the tenth century for Constantine VII Porphyrogenitus, another proof of the fact that agriculture was important for rulers. The work is based on an older collection of the sixth or seventh century. The author of the *Geoponika* is known by the name of Cassianus Bassus. His main sources are Anatolius Vindanius (Vindonius of Berytus) and Didymus of Alexandria, writers of the fourth or fifth century.[21]

---

[20] Verena Winiwarter, "Soil Scientists in Ancient Rome," in *Footprints in the Soil*, pp. 3–16.

[21] A summary and discussion of the sources and history of the *Geoponika* can be found in Robert Rodgers, "Kepopoiia: Garden Making and Garden Culture in

At the beginning of the *Geoponika* the quality of land is discussed: "The best land is that, the soil of which is of a black colour, recommended above all, for it is proof against wet and drought. The next is that of a yellowish hue, and that which is thrown up by rivers, on which they bestow the epithet miry, and that which is sweet, and that which feels warm; for these kinds are known to be adapted to vines and trees, and to the propagation of corn. A deep soil is also recommended, especially if it is friable and not hard to work, and not calculated to the production of trees only; but a red mould is very good for other things, it is not however fit for the production of trees."[22] This short text contains the main indicators for soils in an agricultural context. Besides color and depth, sweetness (that is, the absence of salt and bitterness, both indicators of soils unfit for cultivation) and warmth are listed. Soil quality is seen as being dependent on the plants to be cultivated, not just as an intrinsic quality of the soil.

In the subsequent chapter on soil testing, the wisdom of the ancients is quoted in advising that brackish soil should be avoided. Also, salt is to be avoided in compost and dunghills are to be irrigated with pure and not with brackish water. The twenty-first chapter concerns manure: "Manure makes good land better, and it will be of greater service to that which is bad; but that which is naturally good does not want much manure; that which is of a middling quality, a little more; and that which is thin and weak, a great deal."[23] We further read that manure should be dispersed, as an over-abundance causes scorching. Unmanured land becomes stiff. Compost should not be thrown directly on the roots, but packed between two layers of earth. This will prevent the burning of the roots and, by covering the compost, prevent its heat from evaporating. Heat, one of the qualities discerned in humor theory, can be interpreted in modern terms as the energy contained in the molecules of fertilizing agents such as nitrates, which would indeed be better preserved under a protective layer of earth. Bird droppings, with the exception of that of water birds, are useful, because of their moisture. Pigeon dung is superior, therefore it is sometimes mixed with the seed when planted, without further preparation. In addition to rendering an impotent soil more

Geoponika," in *Byzantine Garden Culture*, ed. Antony Littlewood, Henry Maguire, and Joachim Wolschke-Bulmahn (Washington, D.C., 2002), pp. 159–75.

[22] *Geoponika: Agricultural Pursuits*, trans. Thomas Owen, 2 vols. (London, 1806), vol. 1, p. 54.

[23] *Geoponika*, vol. 1, p. 67.

powerful, pigeon dung can also expel agrostis, a grassy weed, which is mentioned throughout the work. The high appreciation of pigeon dung is shared by Latin writers, but without the explanation that its value is in the moisture it brings.

The *Geoponika* ranks manures according to their quality. Next to that of pigeons, human feces is listed because it is destructive to weeds. Human feces is prepared in Arabia by drying it, then macerating it in water, and drying it again. This manure is used particularly for vines. Third best is donkey's dung, very fertile and of good use for any plant. Goat dung, ranked thereafter, is more pungent than sheep's dung, which comes next, followed by ox dung. Hogs dung is superior in quality, but by its instant heat burns cornfields, and hence is of no use for them. Horse and mule dung is inferior, but can be made useful by mixing it with other sorts. The ranking is slightly different from that of Columella, one of the Latin authors, in that goat's and sheep's droppings are valued higher than cattle excreta by the *Geoponika*.[24] Compost (that is, manure made from dung) should never be used fresh, as this will create many noxious animals (worms). The author(s) think that 3–4 year old compost is best. Finally, the *Geoponika* links the moon's cycle to agricultural operations, again in accordance with Latin writers: manure should never be spread during the waxing of the moon, as this creates more weeds.

Chapter twenty-two of the *Geoponika* is devoted to the preparation of manure. A trench is to be dug in which several kinds of materials are to be mixed: dung of different quality, ashes from the ovens, dirt and human urine, which is considered particularly good for vines. Stubble, when soaked with cattle excrement, becomes a good addition to the manure pit, but it should be used as litter for the cows before adding it. The filth from the currier-yard is the only waste from craft operations mentioned. It should also be added to the trench. Ashes from chaff, thorns, wood and underwood are likewise recommended, as is sea-weed, which has to be washed with sweet water before adding it. The manure in the pit is then liquefied by adding water, and stirring with a pole is recommended. Rain water that is muddy and turbid is particularly good to add, as it will

---

[24] Columella, *De re rustica* 2.14.4, ed. Will Richter, in *Zwölf Bücher über Landwirtschaft: Buch eines Unbekannten über Baumzüchtung*, 3 vols. (München, 1983), vol. 1, p. 186. While this is the best edition and commentary, for quick reference to an English translation, see Columella, *On Agriculture*, trans. Harrison Boyd Ash, 3 vols., The Loeb Classical Library 361, 407 and 408 (Cambridge, MA, 1900).

increase the manure. This mixed manure is never referred to afterwards, a typical feature of the book, probably due to its collective nature. There is no similar recipe in the Latin literature. Interestingly, by mixing animal excreta, which contain mainly nitrogen, ashes from plants, which contain potassium, and specialized items such as sea-weed, which are rich in potassium and, to a lesser extent, in phosphorus or currier-yard remains, which contain about 5-10% nitrogen, but in chemical compounds which are hard to break up, a multi-purpose combination fertilizer could be achieved.[25] Muddy water would have added some soil minerals such as calcium, also one of the main plant nutrients. While there is no way to tell if this preparation was ever made and applied, its mere existence proves the point that by systematic observation, potentially effective prescriptions could be developed. A liquid might have been difficult to transport over longer distances, but it did not need to be plowed under, thus being useful for meadows, vineyards and in arboriculture alike. In the following chapter of the *Geoponika*, the fertilizing treatment is differentiated for different soils. Arabic knowledge or practice is mentioned in several instances, with Arabic terminology used at times.

The authors of the *Geoponika*, like their Roman counterparts, knew that lupines, one of several nitrogen-fixing legumes known in antiquity, make poor ground fertile.[26] They give an elaborate description of how to proceed with lupines planted for the sake of manuring or meliorating the ground. The plants were to be cut before the Ides of May under wet conditions, left to rot on the ground and then ploughed under so that the rotted material would be fully incorporated into the soil, and all roots destroyed.[27] Lupines were believed not only to fertilize the ground, but also to work against weeds. The theory behind the elaborate practice of the *Geoponika* is the aforementioned humor theory, a theory of polarity, according to which the balance of the qualities is healthy or fruitful. In the detailed fifth book on vines, the relation of

[25] While seaweeds are a diverse group of species, they are similar enough with regard to their chemical composition to allow reference to a regional study as being representative of sea-weed in general: D. D. Hong, H. M. Hien and P. N. Son, "Seaweeds from Vietnam used for Functional Food, Medicine and Biofertilizer," *Journal of Applied Phycology* 19 (2007), 817–26, esp. Table 5 (on p. 821). Seaweeds also contain an interesting array of micro-nutrients. Information on currier-yard refuse content can be found in R. Dörr, "Die Verwendung von Leimleder als Stickstoffdünger in der Landwirtschaft," *Landwirtschaftliche Forschung* 15 (1961), 144–50.

[26] *Geoponika*, vol. 1, p. 86.

[27] *Geoponika*, vol. 1, p. 106.

plants and soils is explained explicitly using this theory. Those vines that are naturally moist should be planted in squalid, cold and more dry situations, those that are dry, in moist ones, so the soil will be able to supply what is wanting in the plant.

In the ancient pagan sources of the *Geoponika*, the earth was conceived as a numinous spiritual being, *terra mater*, and its fertility was discussed in similar terminology. For Columella, a first-century pagan author, the farmer was to identify the point in time when the earth as mother showed the greatest willingness to accept the seed and tend it, becoming pregnant with the seed and then needing special care.[28] Such allusions to a numinous goddess are not found in the *Geoponika*, but it remained known that the earth is dynamic with regard to its fertility. At the time of the setting of the star called Stephanos "the ground has a certain natural power of promoting the growth of the seeds that are then sown."[29]

The highest degree of colonization was and is found in gardens. Not only are plant diversity, nutrients and water controlled more extensively in gardens, but garden soils are also subject to a much higher intensity of intervention than others. Soils can be mixed according to the needs of specific plants, they can be mixed with horse dung in hotbeds to force early growth and weeding is much more prevalent in gardens, too. The *Geoponika* gives an overview of land suitable for gardening. Neither a white clay, nor very rough land, nor soil breaking into wide chasms in the summer are useful; white clay that is frozen in winter and dry in the summer destroys everything that is planted in it or makes it weak and small. Even if one mixed it with an equal quantity of manure, it would not be good. Earth that opens in chasms is altogether useless; and that which is rough can neither support the plants, nor does it afford the circulation of water. Some rough and sandy situations are well adapted to esculents, nevertheless. These soils have plenty of nutritive mould, by which the roots are nourished. The *Geoponika* has an interesting description of a test for such a soil. The soil is dissolved in water and washed, and if one finds plenty of nutritive mud, it is considered productive. If it possesses a more watery substance, it is not good for esculent plants. Likewise glutinous mould, soft as wax to the hand, is improper for esculents. What is fascinating about this passage is the fact that soils are dispersed in water for testing. In addition to the

---

[28] For example, Columella, *De re rustica* 10, pp. 140–6 and 161–5.

[29] *Geoponika*, vol. 1, p. 59 and thereafter.

common test for vineyards, where the soil is mixed with water and the liquid is tasted after the settling of the soil particles, this test is for organic matter, which is the mud that does not settle, a quality of soils indirectly perceivable through the soil color without such a test, for which no counterpart exists in the Latin writings.[30] These tests are based on the concept that soil consists of several distinct materials, a conceptually highly important point.

According to the *Geoponika*, gardens are to be manured diligently. Ashes are considered the best manure for esculent plants. Being very small and by nature warm, they kill noxious animals. The second best manure is that of pigeons. Whereas ashes contain mainly potassium, pigeon's dung, like that of the other fowl, contains a high amount of phosphorus, because birds have a different metabolism than mammals. Pigeon's dung has, so the *Geoponika* states, the power to destroy noxious animals and if a little of it is applied, it will produce the same effect as a great quantity of other dung. If phosphorus was the deficient nutrient, which is not totally implausible if green manure such as lupines supplied enough nitrogen, and ashes added potassium, this observation can be explained as an outcome of the Law of the Minimum, which was only formulated in the nineteenth century.[31] The *Geoponika* lists manures for gardens in detail: donkey's dung is considered superior to pigeon's dung by some, as it renders esculent plants more sweet. Goat's dung is also very good. For want of these, other manure can be used, but it should not be fresh, for it produces noxious animals; one-year-old, having been frequently turned over with the spade, is considered better.[32] In the aforementioned case, where a 3–4 year storage period was recommended, most of the nitrogen contained in the compost would have been lost due to volatilization of ammonia or leaching, if the compost was not covered. A one-year delay in manure use would also have led to losses, but not as much. Why would the sources recommend procedures leading to nitrogen loss? Manuring is necessary for the growth of the desired plants, but it provides nutrients also to weeds, that is, undesired plants. Weed seeds are contained in manures. Using

[30] Verena Winiwarter, "Soils in Ancient Roman Agriculture: Analytic Approaches to Invisible Properties," in *Shifting Boundaries of the Real: Making the Invisible Visible*, ed. Helga Nowotny and Martina Weiss (Zurich, 2000), pp. 137–56.

[31] Originally formulated by Carl von Sprengel in 1828 and popularized by Justus von Liebig, the law states that growth is controlled not by the total of resources available, but by the scarcest resource, which is called the limiting factor.

[32] *Geoponika*, vol. 2, p. 95 and thereafter.

fresh manure, even if the nutritive value might be greater, has the disadvantage of increasing the competition by weeds. After several years in the dunghill, most of the weed seeds will have lost their ability to germinate, hence more of the decreased amount of nutrients would benefit the cultivars. The optimum would depend on temperature and weed types and therefore no general rule for the optimal age of manure is possible. Experiences and hence prescriptions in the texts differ.

The *Geoponika* is not a work of great originality. This fact makes it so precious in the context of this stepping-stone approach: it represents what was commonly known to an agricultural writer in the south-east of Europe in the tenth century and is firmly built on the earlier works, thus allowing a glimpse of the tradition. Much has been speculated about the connection of the *Geoponika* to the agricultural writings of Al-Andalus, Arabic Spain. It has been proven that the *Geoponika* reached the west of Europe only via a Persian translation, but the work was influential and is an important part of the multi-lingual and multi-cultural European web of knowledge about soils.[33] The *Geoponika* was translated into German in 1539. This too is typical. It was one of many translations of ancient works made in the fifteenth and sixteenth centuries, before new works based on, but not translated from, the ancient corpus became available, such as Coler's *Oeconomia* published 1596–1606, mentioned here as a German example for many other works published in the vernaculars of early modern Europe.

Let us now move to the very end of the period under consideration to a summary of a submission to the Prussian Academy of Sciences in 1749 by Johann Theodor Eller (1689–1760), entitled *Recherches sur la fertilité de la terre en general.*[34] The piece was published in German in abbreviated form in the *Hamburgisches Magazin*, a more popular venue, two years later. Eller rhetorically bemoans the neglect of agriculture by the literate, educated class, and makes reference to the ancient writers Varro, Columella and Virgil to prove the worthiness of the subject. An important difference between the agricultural experts and Eller, medical doctor and natural scientist, should be noted. While the agriculturalists probe soil qualities in the field, Eller analyzes them in the

---

[33] Julia María Carabaza Bravo, "La Filâha Yûnâniyya et Les Traités Agricoles Arabo-Andalous," *Arabic Sciences and Philosophy* 12 (2002), 155–78.

[34] On Eller's life, see Jörg Heinrich, *Johann Theodor Eller: Ein bedeutender Arzt, Wissenschaftler und Medizinalbeamter in Brandenburg-Preußen in der ersten Hälfte des 18. Jahrhunderts* (Husum, 2003).

laboratory. While they look for indicators, he looks for causes. Eller reports that three different kinds of earth can be discerned according to their reaction upon dissolution in water and subsequent drying. Some soil constituents disperse into dust, some harden. Some are soluble in acid, some are not. The action of fire allows us to differentiate between these classes of material. The first type hardens in fire, so that one can draw fire from the particles. This class consists of clays, fuller's earth, yellowish, fat clays used by brick-makers (loams), white, grindable earths, stone-marrow and others. The second class of materials becomes glassy when heated with alkali. Sands and small gravel belong to this class. The third type of materials shows some initial melting, but breaks into a powder thereafter and becomes calcinated. Chalk, marl, spar, and the ashes of plants and animals belong into this class. Eller builds his theory of soil fertility on these chemical distinctions.

Taking the morphologically-caused diversity of soils into account, Eller holds that the materials of the third class are not part of the "usual" (mineral) constituents: sand, yellow earth for bricks and clay. The alkaline earths of the third class are not part of soils, unless they have been put there by humans in order to increase fertility, because, and this is the basic fertility theory of Eller, they draw the humidity of the air to them. Humor theory and alchemy had gradually lost their predominant role for the explanation of natural phenomena by the end of the seventeenth century, and for a century thereafter, speculative fertility theories based on laboratory experiments gained ground. Eller mentions marl, ashes of plants and animals, and tanning liquor as such materials.

By Eller's time, plant growth had been proven to be possible in sterile sand or in water alone, and soil constituents were no longer thought to be responsible for fertility. Eller reports what he saw as consensus. All natural scientists agreed with Eller that the soil would contribute to the growth of plants only the following: it held the seed and the water around the seed so that germination was fostered, and it also held the roots, so that they could bring humidity to the plant.[35]

In the eighteenth century, when treatises on the improvement of agriculture abounded, economic societies discussed soil fertility and other matters in detail. Can we assess whether Eller's prize-winning essay was

---

[35] J. T. von Eller, "Untersuchungen Von Der Fruchtbarkeit Der Erde Überhaupt," *Hamburgisches Magazin, oder gesammelte Schriften, zum Unterricht und Vergnügen* 8, no. 1 (1751), 230.

in any way typical? A comparison with Krünitz's *Economic Encyclopedia*, a mammoth undertaking in 242 volumes, can help clarify this issue. Krünitz's entry on fertility, published in 1778, offers the famous chemist Wallerius' discussion of fertility in its presentation of the matter.[36] In it, he sums up the controversy. The matter, according to Krünitz, is difficult for several reasons. Some would hold that the uppermost, middle or deep layer of the soil was responsible; others held salt, and in particular nitre, responsible; some believed in manure or in the fattiness of the soil; most would think that the air was responsible. Some considered fire and warmth; many saw water as the reason for fertility. Duhamel de Monceau and Jethro Tull are mentioned as those who considered an easily soluble or easily separable soil as fertile. According to the compiler, something of all of these theories was probably true. Wallerius, however, wanted to differentiate between the material causes, the causes related to form, and the instrumental causes. Krünitz refers to him in detail. The discussion, intended to sum up the knowledge of the time, is in line with Eller, although Krünitz allows for more uncertainty.

The encyclopedia does not only present theory. Krünitz' alphabetical list of materials that can serve as fertilizers is impressive. Ashes, in particular those of soap-making and alkali-production; foliage; human and animal blood; burnt stubble; burnt dried, grassy sods; Duckstein, a mineral used in Lower Saxony; tanning liquor and fuller's hair, all leftovers from the curing of fur; spoilt herring; runch; fusted wood; chippings of horn; hoofs; lime; pine and fir needles; gravel; coal-dust; dirt from the street; chalk; paste from old walls, in particular from sheep or cow-barns; old shingles; burnt glue; woollen rags; cut-up hemp rope; other rags; malt-dust; burnt molehills; marl; moss; mussels or other shells; burnt quitch; burnt or rotted grass; roots of *Brassica rapa*, a type of cabbage; soot, especially sea-coal soot; sawdust; saltpetre; ashes from salt-making, in particular the black ones; leftover liquid used for seed treatment; sand; earth that has been used by animals to scrabble, which is collected from the litter alongside animal paths; schist or shale; debris from old buildings; sea-weed; sea-salt; sea-sand; slurry from the sea; soapy water; pulverized stones; black coal; plowed-in stubble; lake and ditch muck; clay; Toph (the same as Duckstein); peat; plowing in beans, peas, poppy, rapeseed or vetches, the latter when flowering; urine;

[36] D. Johann Georg Krünitz, *Oekonomische Encyklopädie oder allgemeines System der Staats- Stadt- Haus- und Landwirthschaft*, 242 vols. (Berlin, 1776), vol. 15, p. 315 and thereafter.

fuller's earth and pomace. The list contains organic matter, ashes and mineral materials. Taken together, the materials on the list would supply nitrogen, phosphorus, and potassium in varying combinations, as well as calcium, magnesium, carbon and sulphur, thus covering the seven most important plant nutrients.[37]

### *John Evelyn's Terra, 1675*

I offer a stepping stone in-between the tenth and the eighteenth century, a book picked for three reasons. It was written in English and therefore problems of translation can be avoided. It was presented as a scientific work. And it is typical of the literature in its intricate interweaving of old and new knowledge and can stand, much as the *Geoponika*, for the entire genre. John Evelyn's *Terra, A Philosophical Discourse of the Earth* was presented to the Royal Society, of which he was a fellow, in London in 1675. Evelyn was a widely travelled man of independent means with an interest in gardening and various other subjects. He is mostly known for his diaries. His treatise was influenced by the learned discourse of his times, with references both to the ancient sources and to the recent advances of science. Evelyn knew the tradition of knowledge in the field and used it far beyond mere references. Within the discussion of the reasons for fertility, he likens the soil to a womb, a metaphor that had been used since Columella.[38] Eller starts with a list of different soils based on macroscopic and microscopic features, with precise descriptions of different types, gives a list of soil quality indicators based on the ancient literature, and, what is new, sets out to analyze manures microscopically.[39] To Evelyn, the principle of vegetation delivered by the earth is salt, made available to plants by water, the principal agent of growth. Salt to Evelyn is not common salt,

---

[37] For a detailed description with a graphical rendering of nutrient source types of Krünitz's list, see Verena Winiwarter, "'Der Mensch hat Geschichte, weil er die Natur verändert' (M. Godelier): Kulturelles Erbe als Ergebnis von Eingriffen in natürliche Systeme," in *Erb gut? Kulturelles Erbe in Wissenschaft und Gesellschaft*, ed. Ingo Schneider, Margot Schindler and Karl Berger, Referate der 25 Österreichischen Volkskundetagung 2007 in Innsbruck (= Buchreihe der Österreichischen Zeitschrift für Volkskunde N.S.) (2009), forthcoming.

[38] John Evelyn, *Terra. A Philosophical Discourse of Earth, Relating to the Culture and Improvement of it for Vegetation, and the Propagation of Plants, &c. as it was presented to the Royal Society, April 29. 1675* (1675), p. 12.

[39] Evelyn, *Terra*, p. 8.

but the chemical class of substances resulting from the neutralisation of acid and base. He is deeply influenced by alchemical experiments both here and in his consideration that nitre is the manuring salt. Evelyn had read fellow society member Digby's works, whose theory of affinity is built into Evelyn's reasoning as well as *Glauber's Teutschlands Wohlfahrt* (1656). Glauber was an alchemist concerned with nitre, which he saw as the most subtle salt, as some nitrates will, if heated, evaporate explosively. Evelyn incorporated Digby's and Glauber's knowledge into his theory, without being an alchemist himself.

Later in the book, Evelyn discusses manures of different qualities, horse-dung being considered the hottest, donkey's cherished next, for the same reason Columella had given, namely that the donkey's digestion was very complete and therefore the animal produced very refined material. Cattle, sheep, pigeon and pig's excreta are listed thereafter by quality in descending order. Pigeon and other poultry dung is considered to be full of volatile salts, hot, and therefore good for the coldest grounds. Waterfowl were excluded from the list, as their excrement is considered too fiery. Evelyn knows of the power of lupines, regards blood and carrion, ashes, marl, lime, shavings from hair and horn, bones, skins, leather, grape marc and pressings, foliage, saw dust, wood-ashes, wool, sea wrack and pond and ditch muck as fertilizing agents that he differentiates for different uses, with plant-specific recipes for cultivars like artichokes and fig trees. He also relates regional customs, such as fertilizing with sea wrack, which was done on the island of Jersey.[40] Much of what Krünitz would list about a century later can already be found here. Evelyn advises against leaving manure uncovered to the weather, as all its virtue will be lost, and describes the construction of a manure pit in which composts can be mixed to ripen without loss.[41] His book is a remarkable and singular attempt to reconcile traditional experience and the search for a principal cause of vegetation. The idea he put forward, namely that nitrous salts are the main agent of fertility, has not been given due attention in the history of soil fertility knowledge (he is, for example, not even mentioned in Russell's account). Evelyn should be considered as an important node in the web of soil knowledge for his ability to reconcile theory and gardeners' experience, to bring tradition and experiment creatively into contact and context.

---

[40] Evelyn, *Terra*, p. 30 and thereafter. On sea wrack, see p. 34.
[41] Evelyn, *Terra*, p. 38.

## *Conclusions*

Agricultural knowledge about soils was widespread and detailed in pre-modern Europe. Practices to make soil fertile were varied and entailed the use of chemical (manure) and physical (tillage) interventions. In antiquity and the Middle Ages these practices were embedded in a theory we no longer find credible, humor theory, but the concepts of soil constituents and determinants for fertility are nonetheless similar to those employed today. Theory development in early modern times detached itself from these practices in favor of laboratory experiments. Agricultural practices might not have changed as much as the learned literature would suggest, but the pathways to seek new knowledge were certainly shaped by the development of theories of fertility such as those sketched above. Liebig's success in asserting that only mineral nutrients were important for soil fertility would have been implausible, had not the likes of Eller shown that no life-force was needed to explain fertility. What can we learn from this history?

In the tenth century AD, much was already known. In the *Geoponika*, manure is qualified by its heat, and, in the case of bird droppings, by moisture. Manure is considered to be better after its putrid smells have ceased. In addition to dung, ashes, dirt and remains from the curing of leather are mentioned as potentially useful, and elaborate preparations of the dung are mentioned. Manure is matched to soil and to the type of crop. Green manuring with lupines is well known. Salt is considered to be detrimental to fertility. Evelyn, about 800 years later, retains much of this knowledge and adds several more types of manure, mostly other sources of organic matter. His theoretical approach is still based on the theory of the four qualities, as he distinguishes hot and cold grounds and manures. It is different, however, in that he considers volatile salts as the fertilizing principle. Evelyn offers an evidence-based theory, coming from experience, with lots of practical remarks, especially where gardens are concerned. Eller, on the other hand, dismisses practical knowledge. He is interested in a discourse on principles and an analytical approach. Eller dismisses the earth as a fertilizing agent, its ability to capture moisture for him being the only reason why it can act as a fertilizer. Eller is influenced by Van Helmont's experiment published in 1652, or by its repetition by the famous Robert Boyle published in 1661. In both experiments, a tree was grown in a pot. The weight of the soil, the tree and the boiled water they added was recorded meticulously. After five years the plant had gained considerable weight, but the amount of soil was basically the same as it had been when the experiment started, only a few ounces were missing. The tree's weight gain therefore had to come solely from boiled, pure water.[42] The few missing ounces are indeed all that is taken from the earth by plants, but they are vital to plant growth. The Van Helmont experiment was crucial in developing fertility theories. Its misleading interpretation, dismissing the soil as mere growth matrix, was perhaps the most important error in early modern fertility theories, the paradigm shift that prevented later authors like Eller from integrating the 2000 year tradition of knowledge into his framework of thought. Krünitz's encyclopedia shows the division well. Wallerius' fertility theory on the one hand and the impressive and inclusive list of fertilizers on the other hand show soil knowledge after the bifurcation initiated by Van Helmont. The chasm between theory and tradition has been bridged by recent soil science in some ways, but not by reference to earlier writings.

The history of soil knowledge can be used to think about sustainability issues in the context of agriculture, my last point of discussion here. According to Richard Hoffmann, the technical literature on fish cultivation and angling falls into several categories: isolated recipes; unstructured memoranda; brief, organized tracts; and consciously integrated treatises are the main categories for late medieval writings on fishing.[43] Developing his idea further, I would suggest that one can distinguish three types of knowledge within these categories: written reports of practices, practically derived conceptions, and theoretically informed explanations of phenomena that relate to practices. Sustainability research, in the hope of learning from the past, has concentrated on the first type of knowledge, on practices, quite consciously ignoring their context. Traditional ecological knowledge, praised by the International Council for Science in their contribution to the Johannesburg summit, is decontextualized so that the dominant worldview cannot be threatened, much like the *Geoponika* cleansed the soil of its numinous qualities. We are part of a tradition of the re-interpretation of inherited knowledge.[44] Long-term sustainability (sometimes called hard sustainability) cannot be achieved without a major transformation of society. Mazoyer and Roudat in their impressive global overview of the history of agriculture present several examples of societies and their techniques

---

[42] *Russel's Soil Conditions and Plant Growth*, p. 2 and thereafter.
[43] Hoffmann, *Fishers' Craft and Lettered Art*, p. 322.
[44] International Council for Science, 2002.

well-adapted to local circumstances, and thereby potentially more sustainable than "modern," non-adaptive concepts and practices.[45] This points in the same direction as the environmental history of soil knowledge does. As environmental historians, we can contribute to sustainable development by careful, contextualized analysis of past knowledge. It is quite unlikely that we will resurrect humor theory. But knowledge of experience-based, localized, practical concepts of soil fertility and an understanding of its co-creation by colonizing interventions can be of use for organic agriculture as well as for a more general move towards sustainability.

## *Bibliography*

Bergdolt, Ernst. "Beiträge zur Geschichte der Botanik im Orient. I. Ibn Wahhschija: Die Kultur des Veilchens (*Viola odorata* L.) und die Bedingungen des Blühens in der Ruhezeit," *Berichte der Deutschen Botanischen Gesellschaft* 52 (1934), 87–94; reprinted in *Natural Sciences in Islam*, ed. Fuat Sezgin (Frankfurt/Main, 2001), pp. 249–64.

Blum, Winfried E. H. and Verena Winiwarter, "Souls and Soils: A Survey of Worldviews," in *Footprints in the Soil: People and Ideas in Soil History*, ed. Benno Warkentin (Amsterdam, 2006), pp. 107–22.

Bravo, Julia María Carabaza, "La Filâha Yûnâniyya et Les Traités Agricoles Arabo-Andalous," *Arabic Sciences and Philosophy* 12 (2002), 155–78.

Columella, *Zwölf Bücher über Landwirtschaft: Buch eines Unbekannten über Baumzüchtung*, 3 vols. (München, 1983); English translation: Columella, *On Agriculture*, trans. Harrison Boyd Ash, 3 vols., The Loeb Classical Library 361, 407 and 408 (Cambridge, MA, 1900).

de Chadarevian, Soraya. "Laboratory Science Versus Country-House Experiments: The Controversy Between Julius Sachs and Charles Darwin," *The British Journal of the History of Science* 29 (1996), 17–41.

de Montesquieu, Charles. *De l'Esprit de loi* (Geneva, 1748).

Dörr, R. "Die Verwendung von Leimleder als Stickstoffdünger in der Landwirtschaft," *Landwirtschaftliche Forschung* 15 (1961), 144–50.

Evelyn, John. *Terra. A Philosophical Discourse of Earth, Relating to the Culture and Improvement of it for Vegetation, and the Propagation of Plants, &c. as it was presented to the Royal Society, April 29. 1675* (1675).

Feller, Christian, et al., "Charles Darwin, Earthworms and the Natural Sciences: Various Lessons From the Past to Future," *Agriculture, Ecosystems and Environment* 99 (2003), 29–49.

——. "The Principles of Rational Agriculture by Albrecht Daniel Thaer (1752–1828): An Approach to the Sustainability of Cropping Systems at the Beginning of the Nineteenth Century," *Journal of Plant Nutrition and Soil Science* 166 (2003), 687–98.

Fischer-Kowalski, Marina, et al., *Gesellschaftlicher Stoffwechsel und Kolonisierung von Natur: Ein Versuch in Sozialer Ökologie* (Amsterdam, 1997).

*Geoponika: Agricultural Pursuits*, trans. Thomas Owen, 2 vols. (London, 1806).

[45] Marcel Mazoyer and Laurence Roudart, *A History of World Agriculture: From the Neolithic Age to the Current Crisis* (New York, 2005).

Heinrich, Jörg. *Johann Theodor Eller: Ein bedeutender Arzt, Wissenschaftler und Medizinalbeamter in Brandenburg-Preußen in der ersten Hälfte des 18. Jahrhunderts* (Husum, 2003).
Hoffmann, Richard C. *Fishers' Craft and Lettered Art: Tracts on Fishing from the End of the Middle Ages*, Toronto Medieval Texts and Translations (Toronto, 1997).
Hong, D. D., H. M. Hien and P. N. Son, "Seaweeds from Vietnam used for Functional Food, Medicine and Biofertilizer," *Journal of Applied Phycology* 19 (2007), 817–26.
International Council for Science, "Science, Traditional Knowledge and Sustainable Development," ICSU Series on Science for Sustainable Development 4 (Paris, 2002).
Krünitz, D. Johann Georg. *Oekonomische Encyklopädie oder allgemeines System der Staats- Stadt- Haus- und Landwirthschaft*, 242 vols. (Berlin, 1776–1858).
Mazoyer, Marcel and Laurence Roudart, *A History of World Agriculture: From the Neolithic Age to the Current Crisis* (New York, 2005).
McNeill, J. R. and Verena Winiwarter, "Breaking the Sod: Humankind History and Soil," *Science* 304, no. 11 (2004), 1627–9.
Ohlander, Lars, Charlotte Lagerberg, and Ulla Gertsson, "Visions for Ecologically Sound Agricultural Systems," *Journal of Sustainable Agriculture* 14 (1999), 73–9.
Omotayo, Akin and M. W. Musa, "The Role of Indigenous Land Classification and Management Practices in Sustaining Land Use System in the Semi-Arid Zone of Nigeria," *Journal of Sustainable Agriculture* 14 (1999), 49–58.
Patzel, Nikola, Hans Sticher, and Douglas L. Karlen, "Soil Fertility: Phenomenon and Concept," *Journal of Plant Nutrition and Soil Science* 163 (2000), 129–42.
Rodgers, Robert. "Kepopoiia: Garden Making and Garden Culture in Geoponika," in *Byzantine Garden Culture*, ed. Antony Littlewood, Henry Maguire, and Joachim Wolschke-Bulmahn (Washington, D.C., 2002), pp. 159–75.
*Russel's Soil Conditions and Plant Growth*, ed. Alan Wild, 11th ed. (Essex, 1988).
Sieferle, Rolf Peter, and Ulrich Müller-Herold, "Surplus and Survival: Risk, Ruin, and Luxury in the Evolution of Early Forms of Subsistence," *Advances in Human Ecology* 6 (1997), 201–20.
*Soils and Societies: Perspectives from Environmental History*, ed. J. R. McNeill and Verena Winiwarter (Strond, 2006).
von Eller, J. T. "Untersuchungen Von Der Fruchtbarkeit Der Erde Überhaupt," *Hamburgisches Magazin, oder gesammelte Schriften, zum Unterricht und Vergnügen* 8, no. 1 (1751), 230.
Winiwarter, Verena. "Soils in Ancient Roman Agriculture: Analytic Approaches to Invisible Properties," in *Shifting Boundaries of the Real: Making the Invisible Visible*, ed. Helga Nowotny and Martina Weiss (Zurich, 2000), pp. 137–56.
——. "Agrarökologische Geschichtsschreibung: Der Umwelthistorische Beitrag Zur Diskussion Um Nachhaltige Agrar-Entwicklung," *GAIA* 11 (2002), 104–12.
——. "Zwischen Himmel und Erde: Umwelthistorische Aspekte landwirtschaftlicher Kalender," in *Ideologisierte Zeit: Kalender und Zeitvorstellungen im Abendland von der Antike bis zur Neuzeit*, ed. Wolfgang Hameter, Meta Niederkorn-Bruck, and Martin Scheutz, Querschnitte 17 (Innsbruck, 2005), pp. 276–94.
——. "Soil Scientists in Ancient Rome," in *Footprints in the Soil: People and Ideas in Soil History*, ed. Benno Warkentin (Elsevier and Amsterdam, 2006), pp. 3–16.
——. "'Der Mensch hat Geschichte, weil er die Natur verändert' (M. Godelier): Kulturelles Erbe als Ergebnis von Eingriffen in natürliche Systeme," in *Erb gut? Kulturelles Erbe in Wissenschaft und Gesellschaft*, ed. Ingo Schneider, Margot Schindler and Karl Berger, Referate der 25 Österreichischen Volkskundetagung 2007 in Innsbruck (= Buchreihe der Österreichischen Zeitschrift für Volkskunde N.S.) (2009), forthcoming.

# PART TWO

# AQUATIC ECOSYSTEMS AND HUMAN ECONOMIES

# THE SEASONALITY OF FISHING IN MEDIEVAL BRITAIN

Maryanne Kowaleski

The fishing industry and trade of the Middle Ages was subject to a marked seasonality of demand and supply that has mainly been studied as an example of religious influence on diet. The consumption of fish was stimulated by a growing population and increasing urbanization from the eleventh century on, but it was the fasting and abstinence strictures of the Catholic Church that shaped the seasonal demand for fish.[1] By the late Middle Ages, fish was consumed instead of meat on about 40 per cent of the days of the year, including Lent, much of Advent, and the vigils of all holy days, as well as Wednesdays, Fridays and Saturdays.[2] Even more rigorous practices that avoided meat and favored fish were followed by the clergy and particularly devout laity. This essay, however, will focus on the less studied but nevertheless potent influence that seasonality had on the supply of fish, an influence evident in the temporal, geographic, economic, and social organization of the medieval British fishing industry and trade. As early as the eleventh century, the importance of the autumn herring season off the coast of Yarmouth

---

[1] Angelika Lampen, *Fischerei und Fischhandel im Mittelalter: Wirtschafts- und sozialgeschichtliche Untersuchungen nach urkundlichen und archäologischen Quellen des 6. bis 14. Jahrhunderts im Gebiet des Deutschen Reiches* (Husum, Germany, 2000), pp. 37–80; James H. Barrett, Alison M. Locker, and Callum M. Roberts, "'Dark Age Economics' Revisited: The English Fish Bone Evidence AD 600–1600," *Antiquity* 78 (2004), 618–36; and idem, "The Origins of Intensive Marine Fishing in Medieval Europe: The English Evidence," *Proceedings of the Royal Society of London, B series* 271 (2004), 2417–21.

[2] D. Serjeantson and C. M. Woolgar, "Fish Consumption in Medieval England," in *Food in Medieval England: Diet and Nutrition*, ed. C. M. Woolgar, D. Serjeantson, and T. Waldron (Oxford, 2006), pp. 102–30; Christopher Woolgar, "'Take This Penance Now, and Afterwards the Fare will Improve': Seafood and the Late Medieval Diet," in *England's Sea Fisheries: The Commercial Sea Fisheries of England and Wales since 1300*, ed. David J. Starkey, Chris Reid, and Neil Ashcroft (London, 2000), pp. 36–44; Petra J. E. M. Van Dam, "Fish for Feast and Fast: Fish Consumption in the Netherlands in the Late Middle Ages," in *Beyond the Catch: Fisheries of the North Atlantic, the North Sea and the Baltic, 900–1850*, ed. Louis Sicking and Darlene Abreu-Ferreira (The Hague, 2009), pp. 309–36.

compelled the English king to announce a cessation of hostilities at Dover, while one Norwegian invader used "two whole herring seasons" to measure the progress of his seige of Sandwich.[3] In the early days of the Hundred Years War, the king of France was advised to target the autumn herring fishery of Yarmouth because of the huge numbers of English vessels that could be destroyed at one time.[4] The seasonal nature of the herring catch could also affect the timing of marriages, which clustered in the months around Christmas when the herring fleets returned to shore; births in many fishing communities also took on a marked seasonal cast, nine months after the fishers returned home.[5] Seasonality of fishing also prompted the appearance of fishing lodges or cabins to temporarily house fishers and their equipment on otherwise unoccupied beaches during fishing season.[6] Some of these 'cellar settlements' went on to become the nucleus of permanent villages.

The seasonal rhythms of the marine fishing industry were dictated by the migration of fish to their spawning grounds. The best fishing occurs prior to spawning, when the fish are full, plump, and excellent for eating; they also shoal or come together at this time, making it easier to catch large numbers. After spawning, they tend to scatter and their quality declines significantly, although there was a commercial fishery for 'shotten' herring off the Suffolk coast.[7] Most species of fish favor a specific water temperature for spawning; if the climate in a particular year provides too much or not enough sun, then the spawning and egg-hatching process will be affected; the speed and direction of surface winds, fog, and ice can also affect where and when fish migrate.[8] Also

[3] Mark Gardiner, "Shipping and Trade between England and the Continent during the Eleventh Century," *Anglo-Norman Studies* 22 (1999), 82.

[4] Charles de la Roncière, *Histoire de la marine française* (Paris, 1899), pp. 434–5.

[5] These patterns can be measured from the late sixteenth century on, when parish registers survive; see John H. Farrant, "The Rise and Decline of a South Coast Seafaring Town: Brighton, 1550–1750," *Mariner's Mirror* 71 (1985), 61 and 62; and David Butcher, *The Ocean's Gift: Fishing in Lowestoft during the Pre-Industrial Era, 1550–1750* (Norwich, 1995), p. 30. See also Alan Storm, "Seasonality of Births and Marriages in a Seafaring Community before the Age of Steam," *Local Population Studies* 52 (1994), 43–7; and Alain Cabantous, *Les citoyens du large: Les identités maritimes en France (XVIIe–XIXe siècle)* (Paris, 1995), p. 122.

[6] Harold Fox, *The Evolution of the Fishing Village: Landscape and Society along the South Devon Coast, 1086–1550* (Oxford, 2001); and Mark Gardiner, "A Seasonal Fishermen's Settlement at Dungeness, Kent," *Annual Report of the Medieval Settlement Research Group* 11 (1996), 18–20.

[7] *The Bailiffs' Minute Book of Dunwich 1404–1430*, ed. Mark Bailey, Suffolk Record Society, vol. 34 (Woodbridge, 1992), pp. 16 and 145; and Butcher, *Ocean's Gift*, p. 55.

[8] This and the following is a general summary of the major factors that could influence variations in the seasonal migration, abundance, geographic distribution, and size

crucial are conditions that affect the plankton (microscopic algae or animal plankton, including krill) that herring and other pelagic species eat. In the period before they spawn, herring become rich in oil and fat from feeding off plankton; the richer the plankton (which are also highly sensitive to changes in water temperature and salinity), the better the herring catch, so fluctuations from year to year in the size and quality of the catch are heavily influenced by even small changes in the temperature of the water where plankton proliferate. Larger fish such as cod depend on plankton at the larval stage, while the smaller fish and crustaceans that cod and other large fish eat in turn rely on plankton in order to survive and grow, a food chain that stretches all the way down to humans who consume fish. There are, however, no conditions that are ideal for all fish. Herring, cod, and plaice, for example, favor colder temperatures, while hake, pilchard, and red mullet prefer warmer temperatures, so slightly colder water temperatures may yield a larger catch of herring but a dismal return on pilchards off the very same coast.

Herring, the first species to become the target of large-scale commercialized fishing in the Middle Ages, spawn in the British Isles at different times depending on location: in the early spring (particularly in coastal waters, such as off the coast of Ayrshire, the Shetlands, the Hebrides, and Firth of Forth in Scotland, or the northern coast of Ireland), summer (the Irish Sea, and off the coasts of Northumberland, the Shetlands, Hebrides, and north-east Scotland), and winter (in the Straits of Dover, the eastern part of the English Channel, south Devon, and south-east Cornwall).[9] But it was the autumn North Sea herring fisheries stretching from Scotland to Kent that were the most abundant and highly commericalized British fisheries during the Middle Ages.[10]

---

of fish. For analyses of the complexities involved in European marine ecosystems, particularly the influence of climate on fisheries, see, for example, Taivo Laevastu, *Marine Climate, Weather and Fisheries: The Effects of Weather and Climatic Changes on Fisheries and Ocean Resources* (New York and Toronto, 1993); D. H Cushing, *Climate and Fisheries* (New York, 1982); Alister Hardy, *The Open Sea: Its Natural History*, 2 vols. (Boston, 1958–9); Alan Southward, Gerald Boalch and Linda Maddock, "Climatic Change and the Herring and Pilchard Fisheries of Devon and Cornwall," in *Devon's Coastline and Coastal Waters: Aspects of Man's Relationship with the Sea*, ed. David J. Starkey (Exeter, 1988), pp. 33–57; and Jürgen Alheit and Eberhard Hagen, "Long-term Climate Forcing of European Herring and Sardine Populations," *Fisheries Oceanography* 6:2 (1997), 130–9.

[9] For an overview of the different herring fishing seasons in the British Isles, see Cushing, *Climate and Fisheries*, pp. 57–62 and 199–207; Hardy, *Open Sea*, vol. 2, pp. 35–67; and W.C. Hodgson, *The Herring and Its Fishery* (London, 1957), pp. 14–25.

[10] There were other autumn fisheries for herring in Britain off, for example, north Devon, but they commercialized much later.

Starting in mid-summer off the Shetland Islands, one group of herring migrated down the eastern coast of Scotland and northern England by late August and September before moving eastwards towards Norway to over-winter. A second group of North Sea herring migrated from the Northumberland coast down to the Wash (Lincolnshire and northern Norfolk) to spawn in September and October before moving eastwards out to sea for the winter. The third group was the most prolific and the last to spawn, coming very close to the East Anglian coast in October and November and reaching as far south as the English Channel by December and January.

The impact of this seasonal migration is evident in the dates of fairs established in coastal communities along the east coast of England, which were often timed to coincide with the migration of herring past their coasts (Figure 1). Thus the port towns of Northumberland and Durham had a group of August fairs, while further south in Yorkshire the fairs usually began later in August or September, including the ports of Whitby (with a fair on 25 August), Scarborough (15 August-29 September), Filey (23–24 August), and Bridlington (14–15 August). The shoals generally reached southern Yorkshire and northern Lincolnshire by September, which reflects the timing of fairs in Ravenser (7 September-7 October), Grimsby (13–20 September) and Saltfleetby (21–27 September).[11] Visits by foreign fishing craft were also highly seasonal; in early fourteenth-century Scarborough, for example, 85 per cent of the port's trade occurred from early May to mid-October when the herring fishery was active.[12]

Fishers and traders moved south as the herring migrated, but found their greatest catches around Yarmouth in Norfolk from late September through October to mid-November, when the herring shoals moved closer to the shore. The herring catch was so enormous here that a large fair sprang up by the eleventh century; by the early fourteenth

---

[11] For documentary references, see Samantha Letters, *Online Gazetteer of Markets and Fairs in England Wales to 1516* at http://www.history.ac.uk/cmh/gaz/gazweb2.html [accessed June 24, 2007]. Many of these fairs were established very early: the Whitby fair was operating by 1122, Bridlington's was recorded by 1200, Hartlepool's by 1201, Yarmouth's by 1208 (but probably much earlier), and Southwold's by 1223.

[12] Alison Littler, "The Fish in English Economy and Society down to the Reformation" (PhD diss., University of Wales-Swansea, 1979), p. 89; and Wendy R. Childs, "Mercantile Scarborough," in *Medieval Scarborough: Studies in Trade and Civic Life*, ed. David Crouch and Trevor Pearson, Yorkshire Archaeological Society Occasional Paper 1 (Leeds, 2001), pp. 18–22.

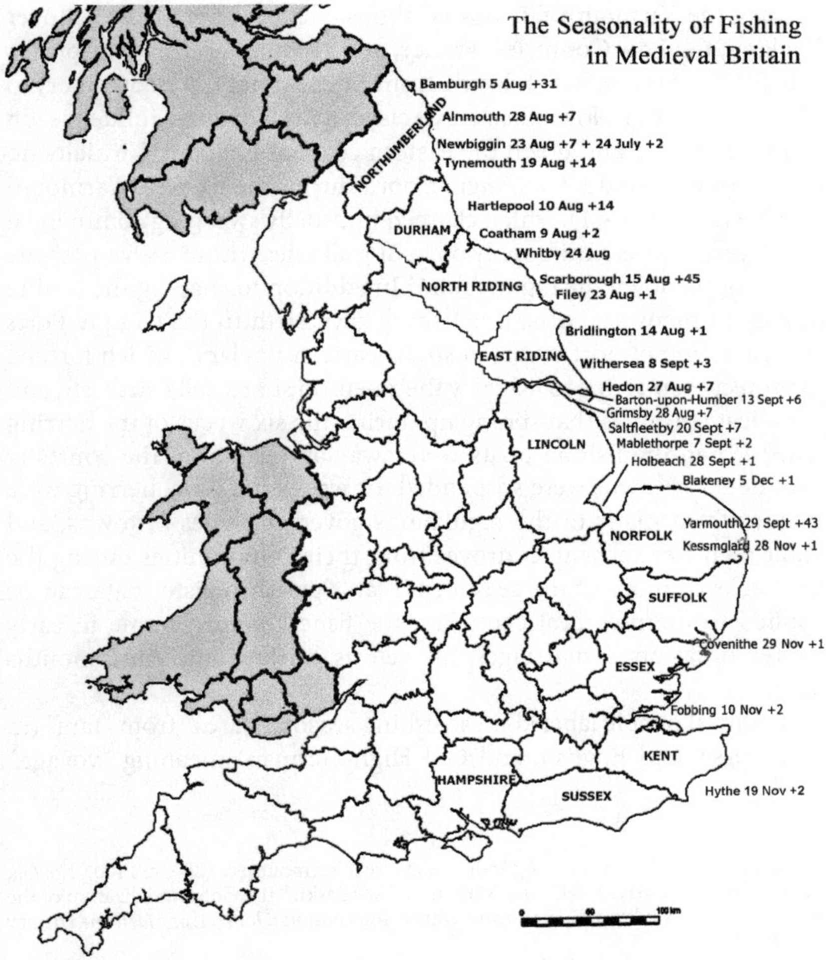


Figure 1. Seasonality of fairs during the North Sea herring season.

*Source*: Samantha Letters: *Online Gazetteer of Markets and Fairs in England to 1516* (http://www.history.ac.uk/cmh/gaz/gazweb2.html, accessed 14 May 2007.) Reproduced from Ordnance Survey map data by permission of the Ordnance Survey © Crown copyright 2001.

century the Yarmouth fair hosted thousands of fishers from all over England, the Low Countries, France, and Germany.[13] This very profitable fishing season, which lasted from Michaelmas (29 September) to Martinmas (mid-November), exercised a tremendous influence on communities up and down the eastern coast of England. Virtually no other fairs occurred at East Anglian ports during the six-week Yarmouth herring fair (Figure 1), which claimed unusually sweeping commercial privileges, such as a monopoly on taxing all sales within a seven-league radius of the town during the fair.[14] In addition to shaping the timing of fairs in many seaports, this fishery gave birth to the Cinque Ports confederation of port towns in south-eastern England, which formed to protect their rights to fish, dry their nets on shore, sell fish freely, and adjudicate disputes that sprang up during the six weeks of the herring fair.[15] When the fishing boats were away at Yarmouth, the courts at Cinque Ports towns were suspended for weeks at a time, herring were exempt from many of the regulations governing sales in towns, and adult men were absent in droves from their communities during the six weeks or more of the season.[16] If later parish-register data can be applied to the medieval centuries, the fishers' return home in early winter occasioned marriages, as well as births some nine months later.

Medieval people labeled these fishing seasons "fares", from "fara" (in Old Norse, Old English, and Old High German) meaning "voyage."

---

[13] Anthony Saul, "The Herring Industry at Great Yarmouth, c. 1280–c. 1400," *Norfolk Archaeology* 38 (1981), 33–43; and Maryanne Kowaleski, "The Commerialization of the Sea Fisheries in Medieval England and Wales," *International Journal of Maritime History* 15 (2003), 185–8.

[14] For the claims made by Yarmouth and the reaction they prompted, see Anthony Saul, "Great Yarmouth in the Fourteenth Century: A Study in Trade, Politics and Society" (D.Phil diss., University of Oxford, 1975), pp. 146–57; and Saul, "Herring Industry," pp. 38–41.

[15] K. M E. Murray, *The Constitutional History of the Cinque Ports* (Manchester, 1935), pp. 17–8. It also gave rise to violent and long-lasting feuds between rival ports seeking to share some of the profits; see F. W. Brooks, 'The Cinque Ports' Feud with Yarmouth in the Thirteenth Century," *Mariners' Mirror* 19 (1933), 27–51; and Saul, "Great Yarmouth in the Fourteenth Century," pp. 146–57.

[16] For sales, see Nathaniel Bacon, *Annals of Ispwich* (Ipswich, 1884), p. 36. For courts, see Dorothy Gardiner, *Historic Haven: The Story of Sandwich* (Derby, 1954), pp. 60–1. For the impact of the fair on particular seaports, see David Sylvester, "The Development of Winchelsea and Its Maritime Economy," in *New Winchelsea Sussex: A Medieval Port Town*, ed. David and Barbara Martin (London, 2004), p. 16. For the following, see n. 5, above.

Fishing fares were often named after their location – hence "Yarmouth fare," "Northlandfare," and "Scarborough fare" – but they could also be named after the type of fish sought, the types of nets used, or the season of the voyage, such as "Winterfare," and "Lentynfare" (Table 1). These labels are found almost exclusively on the east coast of England, where fisheries first commercialized and were more subject to financial exploitation by seigneurial lords and port towns who received a cut of the profits of each fishery, and thus had an incentive to record them in their account rolls. It is common to read, for example, of lords who capitalized on specific fishing seasons. The tenants of Aldeburgh (Suffolk) paid their manorial lord, Snape Priory, 10*s* per year for every boat fishing "in sperling time," while at Rye the fishing dues owed to Fécamp Abbey passed to the Crown when it acquired the town; these dues included "shares" (a percentage of the profits) on fishing at Yarmouth, the winter herring fishery, and Saltfare (probably a deep-sea voyage for cod at Scarborough), as well as the mackerel and plaice summer fishery.[17] Ecclesiastical owners of tithes were no less anxious to insure their portion of profits during particular fishing seasons. The priors and canons of Blythburgh required tithes during "herring, sperlyng fare and *filletfare*" (fishing in shallow waters) from their manor of Walberswick (Suffolk), while Citeaux Abbey, owner of tithes at the great fishing port of Scarborough, negotiated an elaborate agreement with local fishers regarding what percentage of their catch they owed during various fishing seasons, including "in a place called Ferth for taking herring by some in the winter," which was probably near Moray or the Forth in Scotland.[18] In Walberswick, the parish churchwardens also collected a church "dole" or share, and in Dunwich local fishing boats had to pay a town dole of 2*s* while Dutch and other foreign fishers owed 5*s* during

---

[17] W. D. Macray, comp., "Records of the Corporation of Aldebrugh, in the County of Suffolk," *Report of the Royal Commission on Historical Manuscripts*, v. 55, part IV (London, 1907), pp. 300–1; and A. J. F. Dulley, "The Early History of the Rye Fishing Industry," *Sussex Archaeological Collections* 107 (1969), 37–9. For "Saltfare" see also n. 49, below. For the "shares" system in fishing, see Maryanne Kowaleski, "Working at Sea: Maritime Recruitment and Remuneration in Medieval England," in *Ricchezza del mare, ricchezza dal mare. Secoli XIII–XVIII*, Atti delle Settimane di Studi e Altri Convegni, Prato, 11–15 aprile, 2005, ed. S. Cavciocchi (Florence, 2006), pp. 917–22.

[18] *Blythburgh Priory Cartulary*, ed. Christopher Harper-Bill, 2 vols. (Suffolk Records Society, Suffolk Charters, vols. 2–3, 1980–1), vol. 2, pp. 238–40; *Description of Documents contained in the White Vellum Book of the Scarborough Corporation*, ed. I. H. Jeayes (Scarborough, 1914), p. 35; and Peter Heath, "North Sea Fishing in the Fifteenth Century: The Scarborough Fleet," *Northern History* 3 (1968), 56–7.

Table 1. Fishing seasons in some medieval English ports

| Fishing fares | Location | Jan. | Feb. | Mar. | Apr. | May | Jun. | July | Aug. | Sep. | Oct. | Nov. | Dec. |
|---|---|---|---|---|---|---|---|---|---|---|---|---|---|
| | | | | | | | | | | | | | |
| *Scarborough* (Yorks.) | | | | | | | | | | | | | |
| Plaice fishing | Coast | ▒ | ▒ | | | | | | | | | | |
| Lobster fishing | Coast | ▒ | ▒ | ▒ | | | | | | | | | ▒ |
| North Sea fare (cod, haddock) | N. Sea | | █ | █ | █ | | ▒ | ▒ | ▒ | ▒ | ▒ | ▒ | ▒ |
| Icelandfare (cod) | N. Sea | | | | █ | █ | █ | █ | █ | | | | |
| Skate fishing (skate) | Coast/Sea | | | | | | ▒ | ▒ | ▒ | | | | |
| Land herring (herring) | Sea | | | | | | | | | ▒ | ▒ | ▒ | ▒ |
| Winterfare (herring) | Sea | | | | | | | | | | | ▒ ? | ▒ ? |
| | | | | | | | | | | | | | |
| *Dunwich* (Suffolk) | | | | | | | | | | | | | |
| Deyse/Jou fare (?) | Coast | | | | | | | | | | | | |
| Lentynfare (shotten herring) | Sea | ▒ | ▒ | ▒ | | | | | | | | | |
| Cotyldayfare (cod?) | Coast | ▒ ? | ▒ ? | ▒ ? | ▒ ? | ▒ ? | | | | | | | |
| Tramelyfare (various) | Sea/Coast | | | | ▒ | ▒ | ▒ | ▒ | ▒ | ▒ | | | |
| Northlandfare (cod, ling) | N. Sea | | | | █ | █ | █ | █ | █ | | | | |
| Mackerelfare (mackerel) | Sea | | | | | | ▒ | ▒ | ▒ | | | | |
| Flu/Fishingfare (herring) | Sea | | | | | | | | | | ▒ | ▒ | |

| | | | | | | | | | | | | | |
|---|---|---|---|---|---|---|---|---|---|---|---|---|---|
| Fortylfare (herring) | Sea | | | | | | | | | | | | |
| Spratfare (sprats) | Sea | | | | | | | | | | | | |
| Tuckfare (sprats or smelt?) | Coast | | | | | | | | | | | | |
| | | | | | | | | | | | | | |
| *Brighton* (Sussex) | | | | | | | | | | | | | |
| Tucket (plaice) | Coast | | | | | | | | | | | | |
| Shotnet (mackerel) | Sea | | | | | | | | | | | | |
| Drawnet (mackerel) | Coast | | | | | | | | | | | | |
| Scarborough (cod) | N. Sea | | | | | | | | | | | | |
| Harbor (conger) | Sea/Coast | | | | | | | | | | | | |
| Yarmouth (herring) | N. Sea | | | | | | | | | | | | |
| Cock (herring) | Coast | | | | | | | | | | | | |
| Flew (herring) | Sea | | | | | | | | | | | | |

*Sources*: Peter Heath, "North Sea Fishing in the Fifteenth Century: The Scarborough Fleet," *Northern History* 3 (1968), 56–7, 60–1. The North Sea fare was also called "Dogue Drague" (probably a reference to the Dogger Bank cod fishery) and the Winterfare could be termed the "Ferthfare" (probably a reference to Moray or Forth in Scotland); both terms were references to the location of the fishing season. Scarborough's involvement in Icelandic fishing was restricted from about 1412 to the 1430s. *Elizabethan Brighton. The Ancient Customs of Brightelmeston 1580*, eds. Charles Webb and A.E. Wilson (Brighton: Jon Bcal & Son, 1952), 15 20; these customs can be traced back to at least the late 15th century. Mark Bailey, ed. *The Bailiffs' Minute Book of Dunwich 1404–1430* (Suffolk Record Society, vol. 34, 1992), 16–17, 145–46. Fortylfare was carried out in smaller boats called fortyll-boats. The Tramelyfare referred to the use of tramel nets, which could be used to catch a variety of fish, depending on the location and the size of the net mesh. The Deysefare or Jourfare was probably part-time fishing whenever fish were available, most likely done with lines or nets; it might also have involved the use of crab or lobster pots. The Northlandfare was spring and summer fishing for cod and ling off Iceland or the Faroes

*Note*: The darker shading indicates fisheries requiring absences of many weeks and even months.

"the Lenten fare season to Pentecost."[19] In London, the Constable of the Tower of London had the right to tax fishing boats "leaving between Epiphany [6 January] and the Purification of the Blessed Virgin Mary [2 February]" to take small fish in their small nets, as well as "boats called stalbotes from 27 November … to Michaelmas during the season of fishing in Thames Water for fish called sprots."[20]

Further west, from Hampshire up around the southwestern coast through Wales and Cumberland, fishing seasons were not labelled "fares" and were singled out less often, probably a reflection of the later commercialization and the less marked seigneurial exploitation of the western fisheries.[21] The larger range of fish available in western waters also meant that individual fishing seasons were less marked than in the eastern fisheries since good fishing could be had at almost any time of the year – only the type of fish varied. Yet specific fishing seasons were recognized as potentially profitable ventures, as evident in the customary labor services owed by the villein tenants of the south Devon manor of Stokenham. Their lord required them to fish for mullet on alternate days from Candlemas (early February) until the Tuesday after Easter on a specific beach in the manor, where look-out men perched atop rocks on the shore to watch for schools of fish.[22] Once spotted, they signaled down to teams of fishers who rowed out to the location they pointed out, nets at the ready. The lord received one-third of their catch and also had the right to purchase at fixed rates any porpoise, salmon, ling, cod, plaice, bream, ray, and conger they caught throughout the year.

Although the western fisheries developed later, they quickly capitalized on the very rich range of fish species available in their waters,

---

[19] Mark Bailey, "Coastal Fishing off South East Suffolk in the Century after the Black Death," *Proceedings of the Suffolk Institute of Archaeology and History* 37 (1990), 103 and 105–6; *Walberswick Churchwardens Accounts, A.D. 1450–1499* ed. R. W. M. Lewis (London, 1947); BL, Add. Roll 40712 ("*seyssma lentyn far usque Pentecost*"); and *Bailiffs' Minute Book*, ed. Bailey.

[20] The National Archives, Public Record Office [hereafter PRO], E101/4/25, E101/16/31.

[21] For the differing pace of development in the eastern and western fisheries, see Wendy R. Childs, "The Eastern Fisheries," in *England's Sea Fisheries*, ed. Starkey et al, pp. 19–23; Maryanne Kowaleski, "The Western Fisheries," in *England's Sea Fisheries*, ed. Starkey et al., pp. 23–8; and Kowaleski, "Commercialization."

[22] PRO, C134/16/9 (printed in H.P.R. Finberg, "Customs of Stokenham," *Devon and Cornwall Notes and Queries*, 24 [1950], 69–70); and Huntington Library, San Marino, Calif., HAM Box 64/2 (printed in part by William Roberts, "Manorial Stokenham in the Fourteenth Century," in *Stokenham Occasional Papers*, Book 4, ed. William Roberts [1982], 76–97).

particularly off Devon and Cornwall.[23] By the late Middle Ages, there were commercial fisheries in southwestern England for whiting, ling and eel in January; mackerel from February through the early summer and again in the fall after spawning; mullet from February to April and again from August into early autumn; pilchard from July to September; hake from late July until December; herring from the late summer to November; sprats from September through most of the winter; cod and haddock from late October until Christmas; rays in late autumn and early winter; and salmon from late summer into November, to name only a few of the types of fish available in this region. The western fisheries possess all the species found in eastern waters, but they also host species – such as pilchard and hake – that are rarely available in the eastern fisheries. Pilchard in particular is a product associated with Cornwall, where enormous shoals appear off Land's End in July and made their way eastwards down the Cornish coast and along south Devon until late autumn. Because they shoaled so close to the coast during the late summer, they could be taken by seine nets operated by small boats, each rowed by six to eight men directed where to place their nets by a huer who sat high up on shore to spot the location of the shoals.[24]

Hake is another species specific to the western fisheries. A large fish caught by long lines and hooks, hake tend to migrate to their spawning grounds in the Celtic Sea and Bay of Biscay in the first three months of the year, which means their catch rate off the Devon and Cornish coast is highest from February through July; further north and west, it becomes best from August to December, though smaller hake can be caught off Devon and Cornwall into December.[25] The largest hake are

---

[23] For this and the following, see A. J. Southward and G. T Boalch, "The Marine Resources of Devon's Coastal Waters," in *The New Maritime History of Devon, vol. 1: From Early Times to the Late Eighteenth Century*, ed. Michael Duffy et al. (Exeter and London, 1992), pp. 51–60; John C. Bellamy, *The Housekeeper's Guide to the Fish-market for each Month of the Year and an Account of the Fishes and Fisheries of Devon and Cornwall in Respect of Commerce, Economy, Natural History, and Statistics* (London, 1843), pp. 12–34; Maryanne Kowaleski, "The Expansion of the South-Western Fisheries in Late Medieval England," *Economic History Review*, 2nd series, 53 (2000), 429–54; and eadem, "Western Fisheries."

[24] Cyril Noall, *Cornish Seines and Seiners: A History of the Pilchard Fishing Industry* (Truro, 1972), esp. pp. 11–25; and John Scantlebury, "The Development of the Export Trade in Pilchards from Cornwall during the Sixteenth Century," *Journal of the Royal Institution of Cornwall* 10 (1989), 330–6.

[25] Charles Frederick Hickling, *The Hake and the Hake Fishery* (London, 1935), pp. 33–6, 56–8 and 62–3; and John Casey and Javier Pereiro, "European Hake (M. merluccius) in the North-east Atlantic," in *Hake: Biology, Fisheries, Markets*, ed. Jürgen Alheit and Tony J. Pitcher (London and New York, 1995), pp. 125–39.

found further north, in the Shetlands. One sign of the increasing commercialization of the hake fishery was the development in the late fourteenth century of a seasonal pattern that involved the departure of ships in late May and June, many laden with salt, from such southwestern ports as Plymouth and Fowey, and their return from hake fisheries off the southern Irish coast (from Dungarvan around Baltimore to the western coast) in late August and September, with cargoes of lightly salted and partly cured fish.[26] Hake was the prime catch, but conger, ling, and mulwell (cod) was also brought in at this time of year. While away, these English fishers probably anchored in the small bays of south-western Ireland – after negotiating a fee with the local Gaelic chieftans for the right to come ashore for victuals as well as to dry their nets and perhaps salt their catch. Another seasonal Irish fishery entailed departure from the northern ports of Cornwall, Devon, and Somerset in September and a return home in November or December with catches of herring, but also hake, mulwell, and ling.

Long-distance fisheries that involved seasonal absences of two months or more became much more common in the fifteenth century, when English fishers not only increased their participation in the Irish fisheries, but also their activity in other distant fisheries, particularly in Iceland, where cod was the main catch. Cod was available all around the British coasts, but the major European fishing grounds for what some term the "beef of the sea" were off Iceland and in the North Sea near Scandinavia. Most Icelandic and Scandinavian cod reached English tables as stockfish, a wind-dried version that could keep more than a year, though salted cod was also popular. Hull and Bristol were the main English ports involved in Icelandic fishing, but Scarborough and East Anglian ports such as King's Lynn, Dunwich, and Ipswich also participated at one time or another on a fairly large scale.[27] In Hull and Bristol, where the Icelandic trade is particularly well-documented,

---

[26] For this and the following, see Kowaleski, "South-western Fisheries," pp. 430–2 and 442–4; and eadem, "Western Fisheries," pp. 27–8.

[27] For this and the following, see Kowaleski, "Commercialization of Sea Fisheries," pp. 199–200; Wendy Childs, "England's Icelandic Trade in the Fifteenth Century: The Role of Hull," *Northern Seas Yearbook* 5 (1995), 11–31; idem, "Eastern Fisheries," pp. 22–3; idem, "Control, Conflict and International Trade," in *England's Sea Fisheries*, pp. 34–5; E. M. Carus-Wilson, "The Iceland Venture," in *Medieval Merchant Venturers: Collected Studies*, 2nd ed. (London, 1967), 98–142; Heath, "North Sea Fishing in the Fifteenth Century," pp. 57 and 62–4; and G. J. Marcus, "The First English Voyages to Iceland," *Mariner's Mirror* 42 (1956), 313–8. Londoners were also heavily involved, although as merchants and investors rather than fishers.

ships left port in April or May and generally returned by August and September, although the customs accounts largely reflect trading ventures to Iceland since fishermen returning with their own catch were exempt from paying port customs. Most Atlantic cod, however, spawn between January and April, so it is likely that the actual fishing was occurring some months before the traders picked up their cured stockfish. Cod was also the chief catch off Greenland, another long-distance fishery that attracted some English fishers, as well as off Scandinavia, where a few English fishers also ventured.[28] Around 1383, fishermen of Cromer and Blakeney in Norfolk petitioned the king for permission to avoid naval service in order not to miss fishing "to the coasts of Denmark and Norway, and if not then the season will pass and they will be ruined, as will the royal purveyors of fish."[29] Their petition was granted, as long as their ships were under 20 tons. But the largest and most profitable long-distance cod fishery was in North America, particularly the Newfoundland fisheries.[30] The English were probably fishing off the Grand Banks by the last decade of the fifteenth century, but it was not until the late sixteenth century that their involvement became significant and commercially viable. Cod, which spawned from May to September, were caught off Labrador beginning in late June to July, and slightly later off Newfoundland where the season lasted longer.[31]

Unfortunately, we know little about the impact that the long, seasonal absences of fishers had on their families or communities in the

[28] Kirsten A. Seaver, *The Frozen Echo: Greenland and the Exploration of North America, ca. A.D. 1000–1500* (Stanford, 1996), pp. 159–98; Pål Christensen and Alf Ragnar Nielssen, "Norwegian Fisheries 1100–1970. Main Developments," in *The North Atlantic Fisheries, 1100–1976: National Perspectives on a Common Resource*, eds. Poul Holm, D. J. Starkey and Jón Th. Thór, Studia Atlantica 1 (Esbjerg, 1996), pp. 151–2; Carus-Wilson, "Iceland Venture," pp. 195, 117 and 128; and John Brand, *The History and Antiquities of the Town and County of Newcastle upon Tyne*, 2 vols. (London, 1789), vol. 2, pp. 570 and 573.

[29] PRO, SC8/102/5100.

[30] Harold A. Innis, *The Cod Fisheries. The History of an International Economy*, rev. ed. (Toronto, 1954), pp. 11–51; and D. B. Quinn, *England and the Discovery of America, 1481–1620* (New York, 1973), esp pp. 14–6, 20 and 85–6 for the role of early fishers.

[31] Innis, *Cod Fisheries*, pp. 5–8. In the early seventeenth century, when the English presence in Newfoundland was high, ships left England in late March/early April, arrived about five weeks later, began fishing about a month later after preparing their boats and shore-side facilities, fished until the end of July, and returned to England in August, reaching home in September or early October, unless they went first to Mediterranean ports, in which case they would not return to England until late November or December; see Peter E. Pope, *Fish into Wine: The Newfoundland Plantation in the Seventeenth Century* (Chapel Hill and London, 2004), pp. 21–3 and 29.

Middle Ages. But if the timing of early modern marriage and births was shaped in part by the rhythms of the six- to eight-week North Sea herring season, then the two- to four-month absences of fishers plying their craft in Scandinavia, Ireland, Iceland, Greenland and Newfoundland must also have affected the families and communities that were home to large numbers of fishermen. One study of medieval Barcelona, for instance, found that seamen's wives enjoyed unusual legal and inheritance rights that helped families cope with the long absences and early mortality of male kin.[32] Whether this pattern also prevailed in English coastal communities is uncertain, although those living ashore would have had to make some accommodation for women to manage the household and tenancy in the absence of husbands and fathers.[33] They also had to cope with the psychological stress of awaiting the return home of their loved ones from dangerous work in far-flung fisheries. Pirates, heavy seas and violent storms were ever present, evident in the dramatic loss in a sudden gale of twenty-five English ships on their way to Iceland in 1419.[34] Even today, fishing has the highest mortality rates of any occupation.[35] In Elizabethan England, the wives of common seamen often faced considerable financial and psychological hardships while their husbands were away, made worse during periods of increased violence at sea from pirates and privateers.[36]

To these physical dangers were also added significant financial risks since fishing grounds could produce a bountiful harvest one year, and then virtually nothing in other years. These financial risks were compounded in long-distance fishing voyages, which required more capital and labor investment than coastal fishing. In a period without

---

[32] T. M. Vinyoles I Vidal, "La vita quotidiana della gente di mare (esempi barcellonesi dei secoli XIV e XV," *Medioevo: saggi e rassegne* 21 (1996), 9–35.

[33] For evidence of the marked legal, moral, and economic authority of fishers' and mariners' wives in later periods, see Margaret Hunt, "Women and the Fiscal-Imperial State in the Late Seventeenth and Early Eighteenth Centuries," in *A New Imperial History: Culture, Identity, and Modernity in Britain and the Empire, 1660–1840*, ed. Kathleen Wilson (Cambridge, 2004), pp. 29–47, and the essays in *To Work and To Weep: Women in Fishing Economies*, ed. Jane Nadel-Klein and Dona Lee Davis (St. John's, Newfoundland, 1988).

[34] Carus-Wilson, "Iceland Venture," pp. 110 and 119; and Marcus, "First English Voyages to Iceland," p. 315.

[35] For statistics on the occupational dangers facing fishers, see Richard B. Pollnac, "Thresholds of Danger: Perceived Risk in a New England Fishery," *Human Organization* 57:1 (1998), 53–9.

[36] Cheryl A. Fury, *Tides in the Affairs of Men: The Social History of Elizabethan Seamen, 1580–1603* (Westport, 2002), pp. 210–24.

insurance, the loss of capital equipment such as boats, nets, lines or oars in a sea squall or by carelessness was a significant cost from which it would be difficult to recover. These emotional and financial stresses must have grown significantly as long-distance seasonal fishing mounted over the course of the fifteenth century, requiring adjustments from both the mariners and their families who had to cope with their long absences from home. The regular contact with distant ports during these seasonal fisheries could also leave their mark in the fishers' home ports, as in the unwelcome immigration of Irish men and women into the fifteenth-century fishing port of Minehead in north Somerset.[37]

Policy makers and political authorities were also very aware of the seasonality of fishing. Henry IV and other kings of England, for example, issued safe-conducts to French fishers in the midst of the Hundred Years War so that they could participate in the autumn herring fishery; French authorities did the same, giving rise to the *trêves pêcheresses* which became a staple feature of diplomatic relations between England, France, Flanders, and Holland in the fifteenth century.[38] Fishers also sought and were granted exemption from naval service that would interfere with the narrow time frame they had to exploit their fishing grounds.[39] We also see the intervention of crown and local governments to protect fish stocks by outlawing the use of nets that gathered up too many small fry or forbidding fishing altogether when fish were known to breed. Fishers in Rye, for example, were not allowed to fish for plaice with tucknets before Palm Sunday,

---

[37] See, for example, the increasing references to Irish newcomers in the Minehead court rolls, along with fines for harboring Irish vagabonds in Somerset Record Office, DD/L/P26/6, 8, P28/13, 16, P29/9. See also accusations leveled against John Thye, a fisher of Minehead, for carrying home so many "poor and vagabonds from Ireland," in DD/L/P28/13 (court held 8 May 13 Henry VII) and P28/16, fol. 9

[38] *Calendar of Inquisitions Miscellaneous: (Chancery) Preserved in the Public Record Office* [hereafter *CIM*], 8 vols. (London, 1916–2003), vol. 7, p. 313; S. B. J. Noel de la Morinière, *Histoire générale des pêches* (Paris, 1815), pp. 402–6; Michel Mollat, "La pêche à Dieppe au xve siècle," in *Etudes d'histoire maritime (1938–1975)*, ed. Michel Mollat (Torino, 1977), pp. 30–5; Littler, "Fish," pp. 56–8; and *The Red Paper Book of Colchester*, ed. W. Gurney Benham (Colchester, 1902), pp. 109–110. For an order to de-arrest foreign fishing ships, see *Calendar of Close Rolls Preserved in the Public Record Office, Henry III - Henry VII* [hereafter *CCR*], 62 vols. (London, 1892–1975), *1405–09*, p. 157.

[39] See, for example, PRO, SC8/102/5100; SC8/162/8079; *CCR, 1337–1339*, p. 168; *1385–1389*, pp. 42 and 195; and *Calendar of Patent Rolls, Henry III to Henry VII* [hereafter *CPR*], 55 vols. (London, 1891–1916), *1429–1436*, p. 515; *1436–1441*, p. 42; and *1446–1452*, pp. 318–9.

while fines were assessed on a group of Cumberland men who used nets and weirs in the River Eden when fish were known to be breeding.[40] Such prohibitions focused primarily on river and estuarine fishing, where the vulnerability of fish in closed surroundings was more marked, and where royal and local authorities could more easily exercise their authority.

Oysters and salmon were the special target of protective legislation by the government. Oysters bred from April to July along the coast of Essex and Kent where they were safeguarded by both local maritime courts and the admiralty courts, which fined those who dredged for oysters "in prohibited time, with mast and sail, in destruction of the brood [*lez broods*] there, contrary to maritime law and the law of the land."[41] The old saying that oysters should only be eaten when there is an "r" in the month reflects this understanding of the seasonality of the oyster fishery. Restrictions were also put in place regarding the use of dredging instruments – such as the "Wondyrchoun" outlawed in Essex – that destroyed the spat of oysters and mussels as well as depriving the fish of their food by too closely sweeping the mud and seaweed.[42] The central government also went to great lengths to protect salmon – which spawn in shallow, gravelly areas of rivers and streams – by outright prohibitions against fishing during periods when the young fry (known as smolts) were moving downstream in shoals towards the sea. To protect salmon in the Derwent (in Derbyshire) and its tributaries, fishing was forbidden from 1 May to 24 June and from 29 September to 30 November, a period extended from 8 September to 11 November in 1285.[43] The second statute of Westminster (1285) specifically forbade the destruction of young salmon and other fish out of season, but the regular complaints made to Parliament about this problem suggests the

---

[40] *The Records of Rye Corporation: A Catalogue*, ed. Richard F. Dell (Lewes, 1962), p. 66; and *CCR 1396–1399*, p. 437. For other examples, see L. F. Salzman, *English Industries in the Middle Ages* (London, 1964), pp. 272–3; and Childs, "Control, Conflict and International Trade," p. 32.

[41] *Red Paper Book of Colchester*, pp. 123–4. See also Robert H. Goodsall, "Oyster Fisheries on the North Kent Coast," *Archaeologia Cantiana* 80 (1965), 129 and 131; and Patricia Hyde and Duncan Harrington, *Faversham Oyster Fishery through Eleven Centuries* (Folkestone, 2002), pp. 5–8 and 27–8.

[42] "Parliament of January 1377, Text and Translation," in *The Parliament Rolls of Medieval England*, ed. C. Given-Wilson et al. (Leicester, 2005) [hereafter *PROME*], item 50; *PROME*, "Parliament of January 1377, Appendix," item 15; and *CPR, 1374–7*, pp. 489–90.

[43] Salzman, *English Industries*, p. 272; *Statutes of the Realm*, ed. A. Luders et al., 11 vols. (London, 1810–28), vol. 1, pp. 94–5 (c. 47).

statute was not particularly effective.[44] The establishment of weirs and other traps that impeded the migration of salmon in the Thames and Medway Rivers were also a matter of considerable concern, occasioning parliamentary petitions and statute legislation from the thirteenth through fifteenth centuries. In the Parliament of April 1376, for example, a plea was made to remove forever all the nets between London and the sea and to forbid the taking of salmon between Gravesend and Henley bridge on the Thames during spawning season, that is from January to May.[45] The sensitivity of the authorities to the precarious dependency of fishermen on the seasonal migration of fish occasionally extended even to the pope, who responded positively to petitions by fishermen that they be allowed to work on Sundays and holy days during the fishing season.[46]

The economic insecurities inherent in the seasonal fishing that dominated the Middle Ages are evident in the dues (called "shares" in the annual accounts of the bailiffs) collected by the king each month on the profits made by his fishermen-tenants in Winchelsea, a small port town in Sussex.[47] Winchelsea was a major source of fish for the royal household in the thirteenth century, particularly plaice, mackerel, sprats, and whiting, but also sole, conger, dory, haddock, cod, and herring.[48] In the late thirteenth century, the king's shares indicate that the total value of the local inshore fisheries was at least £240 to £400 each year. The more distant fisheries at Yarmouth and Scarborough were even

---

[44] *Statutes of the Realm*, vol. 1, pp. 94–5 (c. 47); vol. 2, pp. 67–8 (c. 19), 89–90 (c. 9); and *PROME*, "Parliament of February 1351," item 28; "Parliament of Feburary 1371," item 18; "Parliament of April 1376," item 56; Parliament of January 1377," item 30; "Parliament of November 1384," item 23; "Parliament of November 1390," item 34; and "Parliament of May 1421," item 16.

[45] *PROME*, "Parliament of April 1376, item 56. Similar petitions about the Thames were heard in "Parliament of January 1377," item 30; "Parliament of November 1384," item 23; "Parliament of November 1390," item 34; "Parliament of September 1402," item 51; and "Parliament of May 1421," item 16.

[46] Salzman, *English Industries*, pp. 273–4; and H. Waquet, "Pecheurs cornouaillais du xv$^{e}$ siècle," *Bulletin de la Société archeologique Finisterre* 40 (1913), 249–50.

[47] This share could vary from ½ to ¼ of one share of the catch, equivalent to "what one sailor gets" and depending on the type of fishery being exploited; for a full discussion, see David G. Sylvester, "Maritime Communities in Pre-Plague England: Winchelsea and the Cinque Ports" (Ph.D. diss., Fordham University, 1999), pp. 233–42. For this and the following, see also Sylvester, "Winchelsea and Its Maritime Economy," p. 16, and Dulley, "Early History of Rye Fishing," pp. 37–41.

[48] *CCR 1247–51*, pp. 54 and 429–30; *1251–53*, p. 68; *1256–59*, p. 153; *Calendar of Liberate Rolls*, 6 vols. (London, 1917–64), vol. 2, p. 127; and Dulley, "Early History of Rye Fishing," pp. 38–9.

more profitable, bringing in roughly £366 and £80 respectively. About 15 boats (with their crews totaling from 60 to 150 men) participated in the Yarmouth autumn herring fishery each year, with about six boats traveling further north to fish off Scarborough for herring and cod during the "Saltfare" in August.[49] At the neighboring port of Rye, herring (caught in the autumn and winter) accounted for 55 per cent of the total fishing dues paid from 1281 to 1285, plaice (sought from April to June) represented 29 per cent, mackerel (fished in the summer) accounted for 10 percent, and sprat only 6 per cent.[50]

Sea erosion forced Edward I to relocate Old Winchelsea in 1280 to New Winchelsea further inland. At first the new town prospered, but as its harbor silted up, fishing profits appear to have declined in the 1290s and the first decade of the fourteenth century, although much of this decrease could have come from the removal of the North Sea fisheries at Yarmouth and Scarborough from the accounts.[51] By the 1340s, the local fisheries were employing about 13 fishers a year, a figure that, not surprisingly, dropped after the Black Death, although in 1371/2, there were 11 active fishers at Winchelsea (Figure 2). Based on the fishing dues collected, there was also a steady drop in the overall profitability of the fisheries, attributable not only to the demographic decline after the Black Death and subsequent plagues, but also the disruption caused by the Hundred Years War. The marked swings in profitability and in the number of fishers – such as the severe decline in the 1360s compared to the early 1350s, but also the difference between 1342/3 and 1344/5 – indicate the volatility of medieval fishing, subject not only to seasonal and annual fluctuations in the quantity of fish available, but also to the vagaries of plague and war (Figure 2).

Although the Winchelsea accounts of fishing shares in the mid-fourteenth century generally depict a period of decline, they are useful for showing the seasonal fluctuations in fishing from month to month (Figure 3). The most intense periods of fishing were in late

---

[49] We are not entirely sure what the "Saltfare" was. Dulley, "Early History of Rye Fishing," p. 38, suggests it was either the Scarborough fishery or the Channel mackerel fishery exploited from April to June. The mackerel fishery, however, was more likely the local fishery reflected in the April peak (Figure 2, below) in the fourteenth-century accounts of shares, which do not seem to include the distant fisheries of Yarmouth and "Saltfare" that were included in the thirteenth-century accounts.

[50] Dulley, "Early History of Rye Fishing," pp. 39 and 41 for this and the following.

[51] Dulley, "Early History of Rye Fishing," p. 41; and Sylvester, "Winchelsea and the Maritime Economy," p. 16.

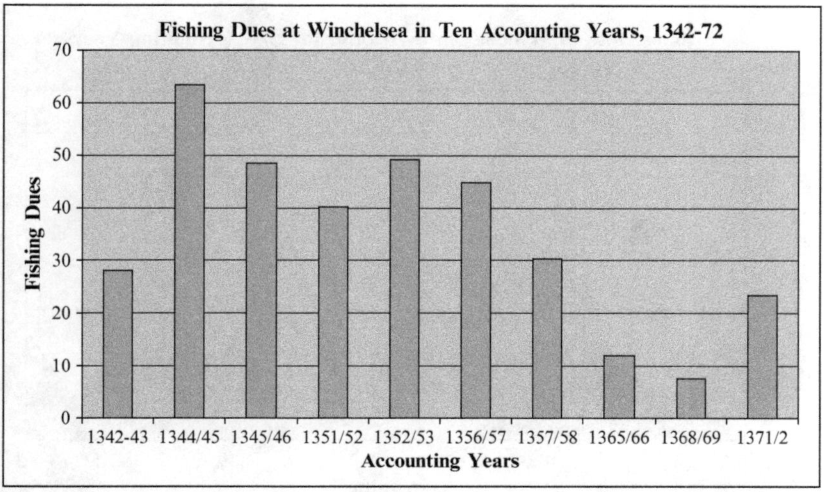


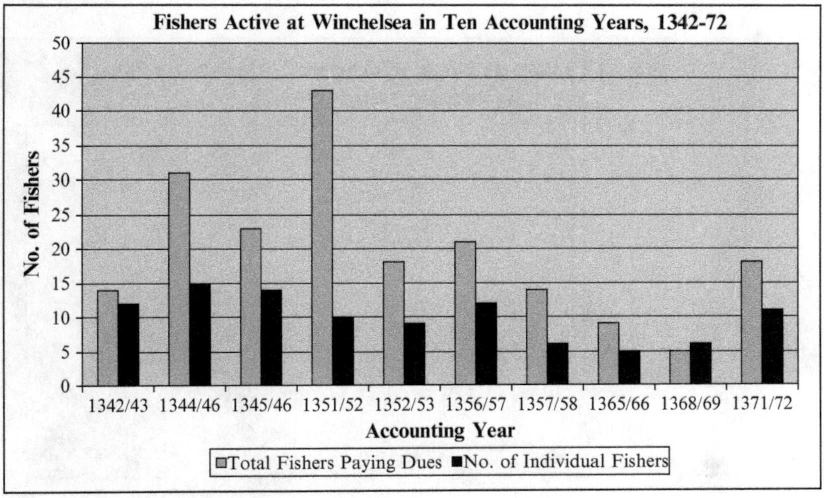


Figure 2. Changes in fishing profits and number of active fishers at Winchelsea in 10 accounting years, 1342–72.

*Source*: PRO, SC6/1032/2 (missing from mid-June to late August), 1032/4, 1032/5, 1032/7, 1032/8, 1032/10, 1032/11, 1032/12, 1032/13.

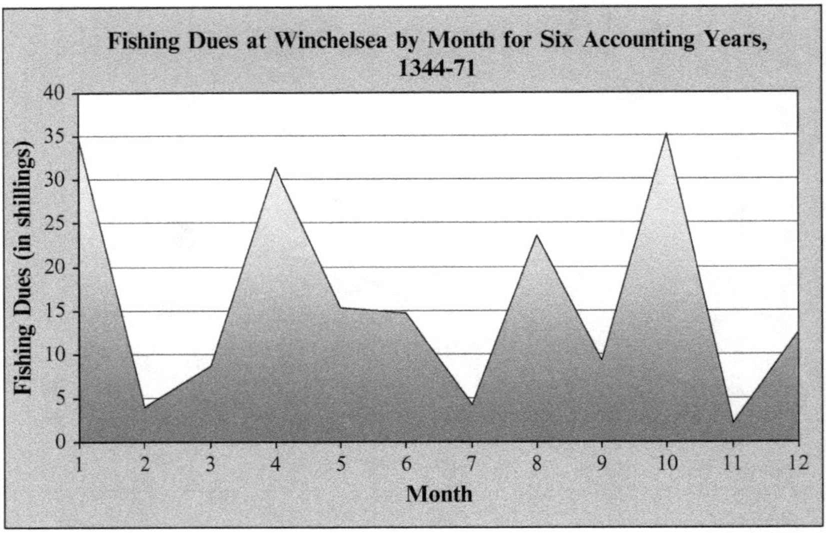


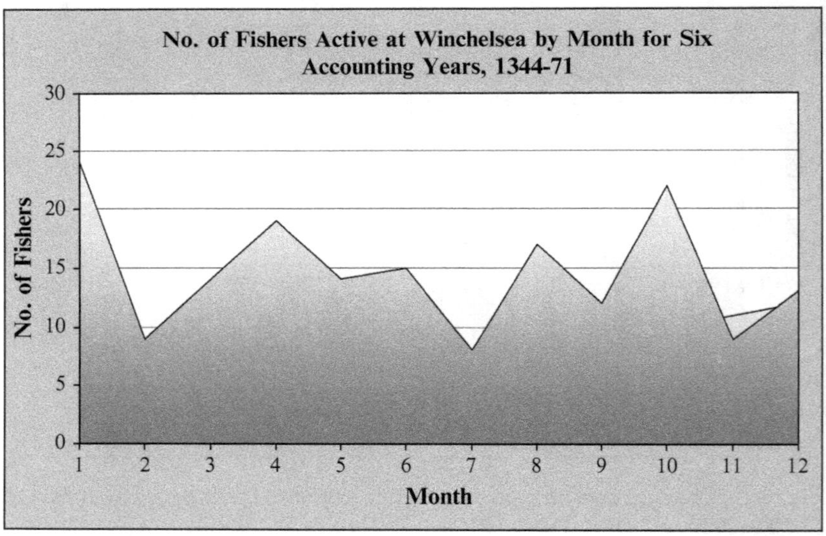


Figure 3. Seasonality of fishing at Winchelsea, 1344–71.

*Source*: PRO, SC6/1032/4, 1032/5, 1032/7, 1032/11, 1032/12, 1032/13.

spring (when plaice were fished, but also mackerel and sole), August (when mackerel came closer to shore but plaice were still fished), October (when local herring approached the shore and whiting were also available), and January, when herring and sprat were sought, but plaice, whiting, conger, sole, skate, shrimp, and cod could also be caught.[52] We are left to speculate about the impact that this seasonal pattern of employment had on the local economy since there were periods – particularly February, July, and November, when activity in the local fisheries was very low. And when we superimpose the pattern of yearly fluctuations (evident in Figure 3) over these monthly oscillations – which left, for example, absolutely no fishers active in eight months of 1342/3 and 1365/6 – we get some idea of the precarious economic situation that fishing seasonality caused for the medieval fishing industry. Similar circumstances existed at other fishing ports, such as Scarborough, where the profits from spring fishing for cod in Iceland plunged between the early and mid-fifteenth century and the returns on summer skate fishing, the autumn herring fishery off Yarmouth, and winter herring fishing off Scotland fluctuated wildly from year to year.[53]

A comparison of the seasonal employment of three Winchelsea fishers demonstrates that individual fishermen did not follow the same seasonal regime from year to year, but made choices about when to put more effort into fishing (Figure 4). All three fishers had their most profitable year in 1344/5, but otherwise their annual income from fishing differed: Alexander Aleyn had his lowest profits in 1345/6, while Bartholomew Bette and Walter Stoket earned less in 1342–3, and Stoket made almost twice as much from fishing as did Aleyn, although the annual variations for Aleyn were much less marked (ranging from 4*s* to 5*s*8*d* to 2*s*10*d*) than for Bartholomew Bette (whose dues ranged from 2*s* to 10*s* to 9*s*). These variations are, however, not that surprising when we consider the different levels of investment or labor that individual fishers put into their work. What is more interesting are the choices these fishers made about when to fish. In 1342–3, Stoket fished in both April and December, but Aleyn worked only in December and Bette

[52] For these fishing seasons, see Dulley, "Early History of Rye Fishing"; J. D. Parry, *Historical Descriptive Account of the Coast of Sussex* (Brighton and London, 1833), pp. 150–2; and L. F. Salzman, "Fisheries," in *The Victoria History of the County of Sussex*, vol. 2, ed. William Page (London, 1907), pp. 264–71.

[53] Heath, "North Sea Fishing."

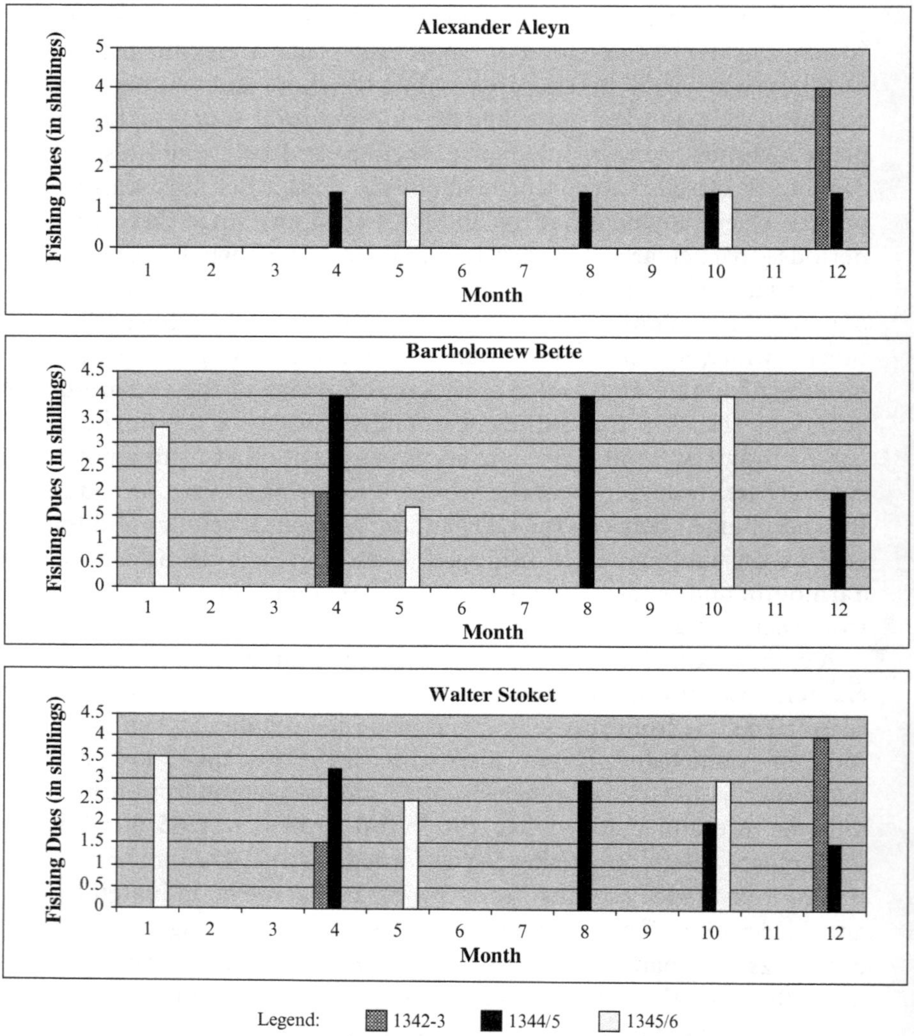


Figure 4. Seasonal employment of fishers at Winchelsea, 1342–46.

*Source*: PRO, SC6/1032/2, 1032/4, 1042/5.

only in April.[54] In the prosperous year of 1344/5, all three fishers chose to work in April, August, and December, but Stoket also fished in October. In 1345/6, all three changed their seasonal pattern by fishing in May and October, although Bette and Stoket also fished in January. In other words, fishers were aware of when the quality of the catch of particular fish made it worthwhile to set out to sea – influenced, most likely, by variations in spawning and schooling patterns that made fishing easier or more plentiful – but individual fishers could also decide to opt out of a particular fishing season altogether. It is impossible to know what factors may have influenced one fisher to skip a season and others to pursue it, but perhaps the suitability of the fisher's equipment for particular catches may have been a factor or simply other claims on the time of individual fishermen, including service on trading vessels or naval ships or even farming.[55]

Although the supply of fish available to medieval consumers was dictated by the seasonal rhythms of the different fish species, the advances made in curing techniques in the late Middle Ages allowed fishermen to respond more directly to seasonal demand, which did not always match seasonal supply. Consumer demand for fish, for example, was at its height during the six weeks of Lent, which generally falls between late February and early April. The intense demand for fish during this long period was responsible in part for Dunwich's "Lentynfare," when shotten herring – normally not at a good stage for eating – became the object of a commercialized fishery (Table 1). The early development of curing processes that made herring and cod last longer also explain why the cod and herring fisheries were the first to commercialize.[56] Herring could be smoked (yielding the famous "red

---

[54] PRO, SC6/1032/2; the fishing dues for 19 June to 28 September are missing from this account (which runs from Michaelmas 1342 to 18 June 1343), which thus misses the mackerel fishing season at its height in August. Note also that by this period the Winchelsea accounts may have been tracking primarily the local fisheries, and not the tenants' activity in the distant (and prosperous) fisheries at Scarborough and Yarmouth; on this point, see Sylvester, "Winchester and Its Maritime Economy," p. 39, esp. n. 88.

[55] For the different types of nets and fishing gear used to catch particular fish, see Dulley, "Early History of the Rye Fishing Industry," pp. 37–48; Heath, "North Sea Fishing," pp. 57–61 and 67–9; Bailey, "Coastal Fishing off South East Suffolk," pp. 103–5; and Table 1, above. For the multiple occupations (including farming) that medieval fishers often practiced, see Kowaleski, "South-western Fisheries," pp. 444–5, 447–8 and 451–2.

[56] Barrett, Locker, and Roberts, "Origins of Intensive Marine Fishing," and idem, "'Dark Age Economics' Revisited."

herring" of Yarmouth) or salted, while cod could be salted and/or wind-dried, the latter method producing the ubiquitous stockfish of the Middle Ages, which reportedly could keep up to two years at a time. By the late fourteenth century, however, new curing processes that involved a light salt cure on board ship, followed (often, but not always) by more intensive curing (salting, brining, smoking, and drying) ashore had begun to take hold in medieval England. These processes played a crucial role in making distant fisheries commercially viable, particularly those in Ireland and Iceland, but also sea fisheries elsewhere, including the English Channel.[57]

The impact of these expanding fisheries that relied on producing more cured fish for the English (and overseas) markets is evident in types of fish brought into the Devon port of Exeter. A series of excellent local port customs accounts allows us to compare all types of fish arriving in the port from the late thirteenth through fifteenth centuries (Figure 5).[58] In the early fourteenth century, the vast majority of fish imported to Exeter (99 per cent of fish cargoes) was herring that arrived from Yarmouth in December and January.[59] This herring reached Exeter bundled in units of one thousand or as lasts (containing 12,000 herring) and only occasionally in cades, a small barrel containing 500 to 1000 herring that had probably undergone extra curing before being barreled. By the late Middle Ages, however, this pattern had changed to reflect the expansion of the south-western fisheries. A far greater proportion of the fish cargoes arrived in February, when Lenten demand was usually peaking (Figure 5); the founding of a new Lenten fair in Exeter in 1374 – which became the most profitable of the town's seven fairs within thirty years – also reflects this response

[57] Richard C. Hoffmann, "Carp, Cods, Connections: New Fisheries in the Medieval European Economy and Environment," in *Animals in Human Histories: The Mirror of Nature and Culture*, ed. Mary J. Henninger-Voss (Rochester, N.Y, 2002), pp. 3–55; Richard Unger, "The Netherlands Herring Fishery in the Late Middle Ages: The False Legend of Willem Beukels of Biervliet" *Viator* 9 (1978), 335–56; Charles L. Cutting, *Fish Saving: A History of Fish Processing from Ancient to Modern Times* (London, 1955), pp. 57–60; Kowaleski, "South-western Fisheries," pp. 438–40, 443, 446–7 and 449–50; and eadem, "Western Fisheries," pp. 26–8.

[58] Most of this fish was cured since fresh fish meant for the direct consumer market was exempt from custom; fresh fish sold to fishmongers for resale was probably liable for custom. For the marketing networks in fish in the Exeter region, see Maryanne Kowaleski, *Local Markets and Regional Trade in Medieval Exeter* (Cambridge, 1995), pp. 307–21.

[59] For this and the following, see also Kowaleski, "South-western Fisheries," pp. 439–40.

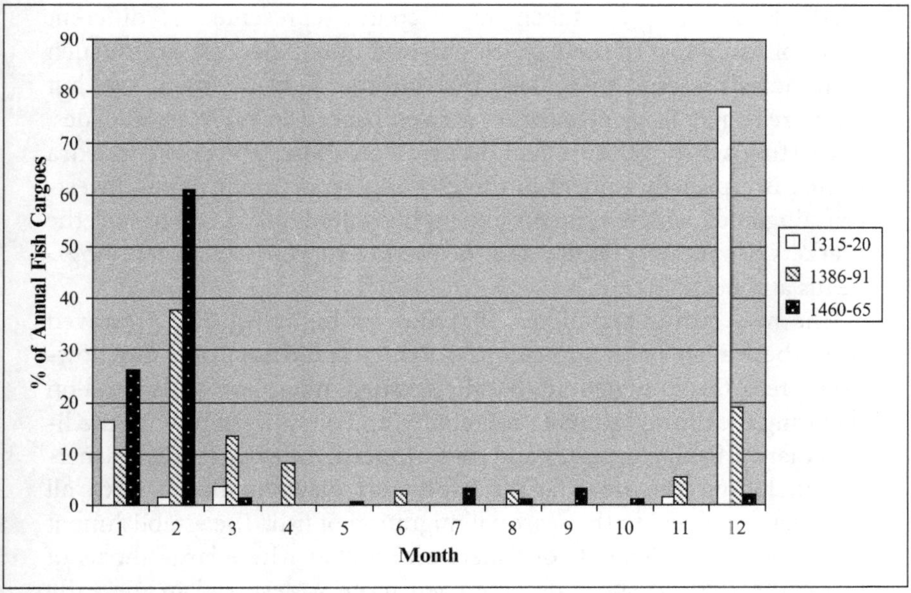


Figure 5. Seasonality of imports of fish to Exeter in three periods, 1315–1465, by percentage of annual fish cargoes.

*Source*: Devon Record Office, Exeter Port Customs Accounts, 1315–20, 1386–91, 1460–64. The accounts for 1315–20 are printed in Maryanne Kowaleski, ed., *The Local Customs Accounts of the Port of Exeter, 1266–1321* (Devon and Cornwall Record Society, n.s. vol. 36, 1993).

to consumer demand.[60] Significant cargoes also came in the summer months, which points to the diversification of the western fisheries as they commercialized their own rich stocks. By the mid-fifteenth century, herring represented barely 20 per cent of imported fish cargoes, which now included 22 varieties of fish that, once different

[60] For the fair, see Kowaleski, *Local Markets and Regional Trade*, pp. 66–7 and 311. Although this new February concentration of fish imports arose to some extent from the expansion of Devon's own herring fishery (January and February were the prime months to catch herring in Bigbury Bay), other fish, particularly hake, actually dominated the February imports at Exeter; for the expansion of this hake fishery, see above. The fish imports from overseas to Devon and Cornish ports (as recorded in PRO, E122/41/4, 201/1 and 115/7 for the 1490s), although a pale shadow of the coastal imports analyzed here, also showed a significant summer distribution by the late fifteenth century, when 58 per cent of overseas fish cargoes arrived in the summer months of July, August, and September; see also Kowaleski, "South-western fisheries," esp. Table 2.

methods of curing are taken into account, represented 41 different types of fish. Most of these species arrived in bundled measures (such as hundreds, seams, or fardels) that indicate light curing at sea, but an increasingly large proportion arrived packed in barrels and cades, suggesting a stronger cure had occurred on shore. The resort to extra curing is especially evident in the growing export trade of southwestern England, which transported highly cured fish – often for the markets of southern France and the Mediterranean – in barrels, hogsheads and butts.

The new curing techniques that allowed higher quality preserved fish to be available over a greater part of the year did not in any way, however, break the stranglehold that the seasonal migration of fish had on dictating the timing, annual profitability, and organization of the medieval marine fishing industry and trade. Indeed, the early modern fisheries, including the new far-off fishery in Newfoundland, were all completely subject to the seasonal migration of fish. The establishment of temporary settlements of fishers on beaches where large shoals of fish came close to shore became even more widespread in the early modern period.[61] And advice written by a West Country merchant in the late sixteenth century counseled merchants to be aware of the delicate timing needed to exploit fully the Scottish and Irish fisheries: " … for the fishing of the owte Ilandes of skotland you must prepare your Shipp that she may be redie to be there shortile After sainte Jamestide for the best fyshinge of cod and Linge is at Mighellmas … "[62]

This essay has touched on the seasonal fisheries of England, but there were equally dominant (and even more commercially important) seasonal fisheries on the Continent, particularly the Scandinavian herring fisheries where in the late fourteenth and early fifteenth centuries, some 17,000 fishers and 8,000 shore-side support workers gathered from late August to early October on the Falsterbo Peninsula at the southern tip of Sweden to catch and process as much as 300,000 barrels of herring in some years.[63] This once vibrant fishery suffered a marked decline

---

[61] See n. 6, above, and H. S. A. Fox, "Cellar Settlements along the South Devon Coastline," in *Seasonal Settlement*, ed. H. S. A. Fox, Vaughan Paper 39, (Leicester, 1996), pp. 61–9.

[62] Printed in N. S. B. Gras, *The Evolution of the English Corn Market from the Twelfth to the Eighteenth Century* (Cambridge, Mass, 1915), p. 430.

[63] A. E. Christensen, "La foire de Scanie," in *La Foire*, Recueils de la Société Jean Bodin 5 (Bruxelles, 1953), pp. 241–66; Poul Holm and Maibritt Bager, "The Danish Fisheries, c. 1450–1800: Medieval and Early Modern Sources and Their Potential for Marine

from the late fifteenth century on, however, for reasons that scholars do not completely understand. Historians have only begun to investigate how the biology of fisheries affected the human organization of marine fisheries in past societies. Climatic changes in particular seem to have dramatically influenced the location, number, and health of spawning fish, as well as the plankton that played such an essential role in feeding fish. Archaeofaunal research has been especially helpful in illuminating these connections, as in recent work on Limfjord in Denmark, where over the course of the Iron and Viking ages plaice and eel were replaced by herring as the entries to the fjord closed off and restricted the flow of highly saline water from the sea.[64] When part of the fjord was once again opened to the sea by a storm in 1624, the flow of very salty North Sea water damaged the herring fisheries and depressed the local fishing industry. Similar investigations that draw on a wide disciplinary range of scholarship need to be done for England and for other parts of Europe. Many questions remain to be addressed, but the work being done by environmental historians such as Richard Hoffmann has pointed the way forward to the interdisciplinary approach, sources, and types of questions that we need to investigate.[65]

## *Bibliography*

*Printed Primary Sources*

Bacon, Nathaniel. *Annals of Ispwich* (Ipswich, 1884).
Bailey, Mark, ed. *The Bailiffs' Minute Book of Dunwich 1404–1430*, Suffolk Record Society, vol. 34 (Woodbridge, 1992).
Benham, W. Gurney, ed. *The Red Paper Book of Colchester* (Colchester, 1902).
*Calendar of Close Rolls Preserved in the Public Record Office, Henry III - Henry VII*, 62 vols. (London, 1892–1975).

Environmental History," in *The Exploited Seas: New Directions for Marine Environmental History*, ed. Poul Holm, Tim D. Smith and David J. Starkey (St John's, Newfoundland, 2001), pp. 108–9. For the seasonal fisheries of northern France, see Jean-Claude Hocquet, "Les pêcheries médiévales," in *Histoire des pêches maritime en France*, ed. Michel Mollat (Toulouse, 1987), pp. 36–129; and Eric Dardel, *La pêche harenguière em France: Étude d'histoire économique et sociale* (Paris, 1941), pp. 32–4.

[64] Inge Bødker Enghoff, "Fishing in the Baltic Region from the 5th Century BC to the 16th Century AD: Evidence from Fish Bones," *Archaeofauna* 8 (1999), 41–85.

[65] For example, see Richard C. Hoffmann, "Medieval Fishing," in *Working with Water in Medieval Europe: Technology and Resource Use*, ed. Paolo Squatriti (Leiden, 2000), pp. 331–93; and idem, "Economic Development and Aquatic Ecosystems in Medieval Europe," *American Historical Review* 101 (1996), 631–69.

*Calendar of Inquisitions Miscellaneous: (Chancery) Preserved in the Public Record Office*, 8 vols. (London, 1916–2003).
*Calendar of Liberate Rolls*, 6 vols. (London, 1917–64).
*Calendar of Patent Rolls, Henry III to Henry VII*, 55 vols. (London, 1891–1916).
Dell, Richard F., ed. *Records of Rye Corporation: A Catalogue* (Lewes, 1962).
Finberg, H. P. R. "Customs of Stokenham," *Devon and Cornwall Notes and Queries* 24 (1950), 69–70.
Given-Wilson, C. et al., ed. *The Parliament Rolls of Medieval England* (Leicester, 2005).
Harper-Bill, Christopher, ed. *Blythburgh Priory Cartulary*, 2 vols. Suffolk Records Society, Suffolk Charters, vols. 2–3, (Woodbridge, 1980–1).
Jeayes, I. H., ed. *Description of Documents contained in the White Vellum Book of the Scarborough Corporation* (Scarborough, 1914).
Lewis, R. W. M. ed. *Walberswick Churchwardens Accounts, A.D. 1450–1499* (London, 1947).
Luders, A. et al, ed., *Statutes of the Realm*, 11 vols. (London, 1810–28).
Macray, W. D., comp. "Records of the Corporation of Aldeburgh, in the County of Suffolk." *Report of the Royal Commission on Historical Manuscripts*, v. 55, part IV (London, 1907), pp. 279–312.

*Secondary Sources*

Alheit, Jürgen and Eberhard Hagen, "Long-term Climate Forcing of European Herring and Sardine Populations," *Fisheries Oceanography* 6 (1997), 130–9.
Bailey, Mark. "Coastal Fishing off South East Suffolk in the Century after the Black Death," *Proceedings of the Suffolk Institute of Archaeology and History* 37 (1990), 102–14.
Barrett, James H., Alison M. Locker, and Callum M. Roberts. "'Dark Age Economics' Revisited: The English Fish Bone Evidence AD 600–1600," *Antiquity* 78 (2004), 618–36.
——. "The Origins of Intensive Marine Fishing in Medieval Europe: The English Evidence," *Proceedings of the Royal Society of London, B series* 271 (2004), 2417–21.
Bellamy, John C. *The Housekeeper's Guide to the Fish-market for each Month of the Year and an Account of the Fishes and Fisheries of Devon and Cornwall in Respect of Commerce, Economy, Natural History, and Statistics* (London, 1843).
Brand, John. *The History and Antiquities of the Town and County of Newcastle upon Tyne*, 2 vols. (London, 1789).
Brooks, F. W. "The Cinque Ports' Feud with Yarmouth in the Thirteenth Century," *Mariners' Mirror* 19 (1933), 27–51.
Butcher, David. *The Ocean's Gift: Fishing in Lowestoft during the Pre-Industrial Era, 1550–1750* (Norwich, 1995).
Cabantous, Alain. *Les citoyens du large: Les identités maritimes en France (XVIIe–XIXe siècle)* (Paris, 1995).
Carus-Wilson, E. M. "The Iceland Venture," in *Medieval Merchant Venturers: Collected Studies*, 2nd ed. (London, 1967), pp. 98–142.
Casey, John and Javier Pereiro, "European Hake (*M. merluccius*) in the North-East Atlantic," in *Hake: Biology, Fisheries, Markets*, ed. Jürgen Alheit and Tony J. Pitcher (London and New York, 1995), pp. 125–47.
Childs, Wendy R. "Mercantile Scarborough," in *Medieval Scarborough: Studies in Trade and Civic Life*, ed. David Crouch and Trevor Pearson, Yorkshire Archaeological Society Occasional Paper 1 (Leeds, 2001), pp. 15–31.
Childs, Wendy R. "The Eastern Fisheries," in *England's Sea Fisheries: The Commerical Sea Fisheries of England and Wales since 1300*, ed. David J. Starkey, Chris Reid, and Neil Ashcroft (London, 2000), pp. 19–23.

Childs, Wendy. "Control, Conflict and International Trade," in *England's Sea Fisheries: The Commerical Sea Fisheries of England and Wales since 1300*, ed. David J. Starkey, Chris Reid, and Neil Ashcroft (London, 2000), pp. 32–35.
Childs, Wendy. "England's Icelandic Trade in the Fifteenth Century: The Role of Hull," *Northern Seas Yearbook* 5 (1995), 11–31.
Christensen, A. E. "La foire de Scanie," in *La Foire*, Recueils de la Société Jean Bodin 5 (Bruxelles, 1953), pp. 241–66.
Christensen, Pål and Alf Ragnar Nielssen. "Norwegian Fisheries 1100–1970. Main Developments," in *The North Atlantic Fisheries, 1100–1976: National Perspectives on a Common Resource*, ed. Poul Holm, D. J. Starkey and Jón Th. Thór, Studia Atlantica 1 (Esbjerg, 1996), pp. 145–68.
Cushing, D. H. *Climate and Fisheries* (New York, 1982).
Cutting, Charles L. *Fish Saving: A History of Fish Processing from Ancient to Modern Times* (London, 1955).
Dardel, Eric. *La pêche harenguière en France: Étude d'histoire économique et sociale* (Paris, 1941).
Dulley, A. J. F. "The Early History of the Rye Fishing Industry," *Sussex Archaeological Collections* 107 (1969), 36–64.
Enghoff, Inge Bødker. "Fishing in the Baltic Region from the 5th Century BC to the 16th Century AD: Evidence from Fish Bones," *Archaeofauna* 8 (1999), 41–85.
Farrant, John H. "The Rise and Decline of a South Coast Seafaring Town: Brighton, 1550–1750," *Mariner's Mirror* 71 (1985), 59–76.
Fox, H. S. A. "Cellar Settlements Along the South Devon Coastline," in *Seasonal Settlement*, ed. H. S. A. Fox, Vaughan Paper 39 (Leicester, 1996), pp. 61–9.
Fox, Harold. *The Evolution of the Fishing Village: Landscape and Society along the South Devon Coast, 1086–1550* (Oxford, 2001).
Fury, Cheryl A. *Tides in the Affairs of Men: The Social History of Elizabethan Seamen, 1580–1603* (Westport, Conn., 2002).
Gardiner, Dorothy. *Historic Haven: The Story of Sandwich* (Derby, 1954).
Gardiner, Mark. "A Seasonal Fishermen's Settlement at Dungeness, Kent," *Annual Report of the Medieval Settlement Research Group* 11 (1996), 18–20.
——. "Shipping and Trade between England and the Continent during the Eleventh Century," *Anglo-Norman Studies* 22 (1999), 71–93.
Goodsall, Robert H. "Oyster Fisheries on the North Kent Coast," *Archaeologia Cantiana* 80 (1965), 118–51.
Gras, N. S. B. *The Evolution of the English Corn Market from the Twelfth to the Eighteenth Century* (Cambridge, Mass., 1915).
Hardy, Alister. *The Open Sea: Its Natural History*, 2 vols. (Boston, 1958–9).
Heath, Peter. "North Sea Fishing in the Fifteenth Century: The Scarborough Fleet," *Northern History* 3 (1968), 53–69.
Hickling, Frederick Charles. *The Hake and the Hake Fishery* (London, 1935).
Hocquet, Jean-Claude. "Les pêcheries médiévales," in *Histoire des pêches maritimes en France*, ed. Michel Mollat (Toulouse, 1987), pp. 36–129.
Hodgson, W. C. *The Herring and Its Fishery* (London, 1957).
Hoffmann, Richard C. "Carp, Cods, Connections: New Fisheries in the Medieval European Economy and Environment," in *Animals in Human Histories: The Mirror of Nature and Culture*, ed. Mary J. Henninger-Voss (Rochester, NY, 2002), pp. 3–55.
——. "Economic Development and Aquatic Ecosystems in Medieval Europe," *American Historical Review* 101 (1996), 631–69.
——. "Medieval Fishing," in *Working with Water in Medieval Europe: Technology and Resource Use*, ed. Paolo Squatriti (Leiden, 2000), pp. 331–93.
Holm, Poul and Maibritt Bager, "The Danish Fisheries, c. 1450–1800: Medieval and Early Modern Sources and Their Potential for Marine Environmental History," in *The Exploited Seas: New Directions for Marine Environmental History*, ed.

Poul Holm, Tim D. Smith and David J. Starkey (St John's, Newfoundland, 2001), pp. 97–102.
Hunt, Margaret. "Women and the Fiscal-Imperial State in the Late Seventeenth and Early Eighteenth Centuries," in *A New Imperial History: Culture, Identity, and Modernity in Britain and the Empire, 1660–1840*, ed. Kathleen Wilson (Cambridge, 2004), pp. 29–47.
Hyde, Patricia and Duncan Harrington. *Faversham Oyster Fishery through Eleven Centuries* (Folkestone, 2002).
Innis, Harold A. *The Cod Fisheries. The History of an International Economy*, rev. ed. (Toronto, 1954).
Kowaleski, Maryanne. *Local Markets and Regional Trade in Medieval Exeter* (Cambridge, 1995).
——. "The Commercialization of the Sea Fisheries in Medieval England and Wales," *International Journal of Maritime History* 15 (2003), 180–231.
——. "The Expansion of the South-Western Fisheries in Late Medieval England," *Economic History Review*, 2nd series, 53 (2000), 429–54.
——. "The Western Fisheries," in *England's Sea Fisheries: The Commerical Sea Fisheries of England and Wales since 1300*, ed. David J. Starkey Chris Reid, and Neil Ashcroft (London, 2000), pp. 23–8.
——. "Working at Sea: Maritime Recruitment and Remuneration in Medieval England," in *Ricchezza del mare, ricchezza dal mare: Secoli XIII–XVIII*, ed. S. Cavciocchi, Atti delle Settimane di Studi e Altri Convegni, Prato, 11–15 aprile, 2005 (Florence, 2006), pp. 907–36.
Lampen, Angelika. *Fischerei und Fischhandel im Mittelalter: Wirtschafts- und sozialgeschichtliche Untersuchungen nach urkundlichen und archäologischen Quellen des 6. bis 14. Jahrhunderts im Gebiet des Deutschen Reiches* (Husum, Germany, 2000).
Laevastu, Taivo. *Marine Climate, Weather and Fisheries: The Effects of Weather and Climatic Changes on Fisheries and Ocean Resources* (New York and Toronto, 1993).
Letters, Samantha. *Online Gazetteer of Markets and Fairs in England Wales to 1516*, at http://www.history.ac.uk/cmh/gaz/gazweb2.html.
Littler, A. "The Fish in English Economy and Society down to the Reformation," (Ph.D. diss., University of Wales-Swansea, 1979).
Marcus, G. J. "The First English Voyages to Iceland," *Mariner's Mirror* 42 (1956), 313–18.
Mollat, Michel. "La pêche à Dieppe au xve siècle," in *Etudes d'histoire maritime (1938–1975)*, ed. Michel Mollat (Torino, 1977), pp. 1–43; first published in *Bulletin de la Société d'Emulation de Rouen et de la Seine Inférieure* (1938), 169–211.
Murray, K. M. E. *The Constitutional History of the Cinque Ports* (Manchester, 1935).
Nadel-Klein, Jane and Dona Lee Davis, ed. *To Work and To Weep: Women in Fishing Economies* (St. John's, Newfoundland, 1988).
Noall, Cyril. *Cornish Seines and Seiners: A History of the Pilchard Fishing Industry* (Truro, 1972).
Noel de la Morinière, S. B. J. *Histoire générale des pêches* (Paris, 1815).
Parry, J. D. *Historical Descriptive Account of the Coast of Sussex* (Brighton and London, 1833).
Pollnac, Richard B. "Thresholds of Danger: Perceived Risk in a New England Fishery," *Human Organization* 57:1 (1998), 53–9.
Pope, Peter E. *Fish into Wine: The Newfoundland Plantation in the Seventeenth Century* (Chapel Hill and London, 2004).
Quinn, D. B. *England and the Discovery of America, 1481–1620* (New York, 1974).
Roberts, William. "Manorial Stokenham in the Fourteenth Century," *Stokenham Occasional Papers*, Book 4 (1982), 76–97.
Roncière, Charles de la. *Histoire de la marine française* (Paris, 1899).

Salzman, L. F. "Fisheries," in *The Victoria History of the County of Sussex*, ed. William Page, vol. 2 (London, 1907), pp. 264–71.
Salzman, L. F. *English Industries in the Middle Ages*, new ed. (London, 1964).
Saul, Anthony. "Great Yarmouth in the Fourteenth Century: A Study in Trade, Politics and Society," (D.Phil diss., University of Oxford, 1975).
Saul, Anthony. "The Herring Industry at Great Yarmouth, c. 1280–c. 1400," *Norfolk Archaeology* 38 (1981), 33–43.
Scantlebury, John. "The Development of the Export Trade in Pilchards from Cornwall during the Sixteenth Century," *Journal of the Royal Institution of Cornwall* 10 (1989), 330–59.
Seaver, Kirsten A. *The Frozen Echo: Greenland and the Exploration of North America, ca. A. D. 1000–1500* (Stanford, 1996).
Serjeantson, D. and C.M. Woolgar. "Fish Consumption in Medieval England," in *Food in Medieval England: Diet and Nutrition*, ed. C. M. Woolgar, D. Serjeantson, and T. Waldron (Oxford, 2006), pp. 102–30.
Southward, A. J. and G. T Boalch. "The Marine Resources of Devon's Coastal Waters," in *The New Maritime History of Devon, vol. 1: From Early Times to the Late Eighteenth Century*, ed. Michael Duffy et al., (Exeter and London, 1992), pp. 51–61.
Southward, Alan, Gerald Boalch and Linda Maddock. "Climatic Change and the Herring and Pilchard Fisheries of Devon and Cornwall," in *Devon's Coastline and Coastal Waters: Aspects of Man's Relationship with the Sea*, ed. David J. Starkey (Exeter, 1988), pp. 33–57.
Storm, Alan. "Seasonality of Births and Baptisms in a Seafaring Community before the Age of Steam," *Local Population Studies* 52 (1994), 43–7.
Sylvester, David G. "Maritime Communities in Pre-Plague England: Winchelsea and the Cinque Ports," (PhD diss., Fordham University, 1999).
——. "The Development of Winchelsea and Its Maritime Economy," in *New Winchelsea Sussex: A Medieval Port Town*, ed. David and Barbara Martin, Institute of Archaeology, University College London, Field Archaeology Unit Monograph 2 (London, 2004), pp. 7–19.
Unger, Richard. "The Netherlands Herring Fishery in the Late Middle Ages: The False Legend of Willem Beukels of Biervliet," *Viator* 9 (1978), 335–56.
Van Dam, Petra J. E. M. "Fish for Feast and Fast: Fish Consumption in the Netherlands in the Late Middle Ages," in *Beyond the Catch: Fisheries of the North Atlantic, the North Sea and the Baltic, 900–1850*, ed. Louis Sicking and Darlene Abreu-Ferreira (Leiden, 2009), pp. 309–36.
Vinyoles I Vidal, T. M. "La vita quotidiana della gente di mare (esempi barcellonesi dei secoli XIV e XV," *Medioevo: saggi e rassegne* 21 (1996), 9–35.
Waquet, H. "Pecheurs cornouaillais du xve siècle," *Bulletin de la Société archeologique Finisterre* 40 (1913), 249–50.
Woolgar, Christopher. "'Take This Penance Now, and Afterwards the Fare will Improve': Seafood and the Late Medieval Diet," in *England's Sea Fisheries: The Commercial Sea Fisheries of England and Wales since 1300*, ed. David J. Starkey, Chris Reid, and Neil Ashcroft (London, 2000), pp. 36–44.

# REELING IN THE EELS AT LA TRINQUETAILLE NEAR ARLES*

Constance H. Berman

While records preserved by the Order of the Hospitallers are usually associated with histories of crusade and reconquest, the military religious orders also had many properties and small communities in the west as well. The archival documents about land-holding that these western communities produced, much like those created by other monastic and clerical groups of the time, can have considerable interest for western European history. This is particularly the case for those communities located in areas like the lower Rhône river valley and its delta, the Camargue. It is possible that surviving records for the military religious Orders are thicker on the ground for this than for any other single region in western Europe.[1] It was here at Aigues-Mortes that Louis IX (1214–1270) built the embarkation port for his crusade.[2] Both Templars and Hospitallers had major headquarters at Saint-Gilles and the Hospitallers in support of the crusade would engulf many earlier pious foundations made for the care of local populations into the Order of the Hospital of Saint John of Jerusalem. Such Hospitaller foundations in the Camargue and lower Rhône valley found places in the interstices between other religious foundations of the region because the delta region was not only an embarkation point to the East, but a major internal frontier. The regrouping of peasant populations from more isolated areas into the vicinity of towns and castles, which has been called *incastellamento*, had begun in the eleventh century to create real villages for the first time, and this opened up rural spaces for

---

* This paper, in taking up the evidence for eels on the lower Rhône, is in appreciation of the many ways in which Richard Hoffmann has counseled and collaborated with me. I would also like to thank Richard Unger who long ago encouraged me to look at the la Trinquetaille cartulary and Scott G. Bruce for useful comments and bibliography.

[1] Damien Carras, *L'Ordre du Temple dans la basse Vallée du Rhône (1124–1312): Ordres militaries, croisades et société méridionales*, Collection d'histoire et d'archéologie médiévales 17 (Lyon, 2005).

[2] William Chester Jordan, *Louis IX and the Challenge of the Crusade* (Princeton, 1979), pp. 71–6.

new monastic, religious, or military-religious groups to establish themselves not so much by their own efforts at clearance and reclamation as by profiting from the efforts of their anonymous predecessors.[3] The constantly changing dimensions of landscape, fields, river beds, ditches and channels, levees, salt and fresh water marshes, woodland, pasture, and salt pans created a fluidity of land ownership and political control in a highly productive region which could provide material for the crusading effort. The Hospitaller records offer tantalizing hints of efforts to supply timber, grain, salt, and meat and fish in brine or salt to the East or to Spain for the Christian armies on the major military fronts of the twelfth and thirteenth centuries. But these documents also provide information on local habitats and changes in access to fisheries, particularly with regard to *Anguilla anguilla*, the European eel.

This paper will discuss such hints of habitat change found in documents surviving in a book of charters, the cartulary of the medieval hospital at Arles of Saint-Thomas of la Trinquetaille. La Trinquetaille was part of the system of commanderies and dependencies of the great Hospitaller commandery at nearby Saint-Gilles. Its documents, along with associated sources for the important Porcelet family of Arles, provide considerable evidence for the rapidly changing medieval ecology in the lower Rhône valley and its delta, the Camargue. Changes in the environment itself resulted in changes in patterns of exploitation and water use, documented as access conflicts and fears of over-exploitation of eel fisheries in the medieval Camargue, an economy which continued until extensive pollution in the lower Rhône and Camargue in our own lifetimes threatened those fisheries.

Records of disputes over waterlands also have bearing on the environmental history of the region where the Rhône's fresh water reaches the sea. These sources suggest that the expansion of drainage, the construction of levees and channels, and changes in the riverbed were a result of both natural and human activities, creating an environment in which eel fishing became a major resource, and in which medieval eel farming may have been possible. For those activities to prosper, they required close control by village communities and authorities to prevent overfishing. Documents also mention payments in eels to local religious

[3] Constance H. Berman, *Medieval Agriculture, the Southern-French Countryside, and the Early Cistercians: A Study of Forty-Three Monasteries* (Philadelphia, 1986), pp. 72–4; and eadem, "Construction de moulins, défrichements et crues au onzième et douzième siècles dans le bassin de la Garonne," *Moulins de France* 73 (January, 2008), 6–11.

communities, reminding us of their utility both to monks and to those cared for by new hospitals. The Hospitaller Order may have possibly provisioned crusaders with salted eels. This frontier territory, changing its shape with each season, with each new construction, with rights to transport on its waterways and salt production still up for grabs, was also a territory particularly well suited for certain stages in the development of eel populations. I begin with the hospital of la Trinquetaille and its relationship to Saint-Gilles before turning to what documents reveal about the riverine environment and the effects of human intervention in natural systems, particularly with respect to the harvest of eels.[4]

The hospital of Saint Thomas at la Trinquetaille was located in a suburb of the city of Arles on the right bank of the Rhône River, near what is today the bridge at la Trinquetaille. Its history has been somewhat overshadowed by that of the nearby Hospitaller priory of Saint-Gilles, which also served as the main commandery for the Order in this region. Nonetheless, La Trinquetaille was an independent hospital with its own endowment and, like the priory at Saint-Gilles, with its own medieval book of charters, or cartulary.[5] Like many hospice or hospital foundations made during this period, la Trinquetaille's origins and original purpose are obscure. There is no foundation charter or named founder. We have no evidence for how it came to be associated in the late 1120s

[4] What we know today about eel ecology is not much more than was apparent in the Middle Ages. Both *Anguilla anguilla*, the European eel, and *Anguilla rostrata*, the American eel, hatch at sea, possibly in both cases in that region off Bermuda called the Saragossa Sea. They then develop into eelings or elvers, migrating towards coastlines, where they are found in briny areas: muddy estuaries, tidal marshes, stream mouths and harbors. Eventually they move upstream into fresh water, possibly when rivers become warmer in the spring. Maturing over some years, they return to their mid-Atlantic grounds to spawn and die. Studies of the eel habitat at the Êtang de Vaccarès, just south of Arles, are currently being conducted at Tour du Vallat, a research center for the conversation of Mediterranean wetlands in the heart of the Camargue. They have revealed the importance of eel fishing in that region until the recent decline in their numbers due to pollution. See http://en.tourduvalat.org

[5] See the introduction of *Cartulaire de la Trinquetaille*, ed. Paul A. Amargier (Gap, 1972) for details about the foundation and its ties to the Order. The church of Saint Thomas appears as early as 1110 and the hospital associated with it by the 1130s. That of Saint-Gilles may date from about 1112: *Cartulaire du Prieuré de Saint-Gilles de l'Hôpital de Saint-Jean de Jérusalem (1129–1210)*, ed. Daniel le Blévec and Alain Venturin (Paris, 1997). See also Dominic Selwood, *Knights of the Cloister: Templars and Hospitallers in Central-Southern Occitania 1100–1300* (Woodbridge, 1999). On the organization that had developed for these houses by the late twelfth century, see Judith Bronstein, *The Hospitallers and the Holy Land: Financing the Latin East, 1187–1274* (Woodbridge, 2005). Carraz, *L'Ordre du Temple*, pp. 103–4, treats la Trinquetaille as equal to Saint-Gilles.

with the brothers of the Hospital of the Holy Sepulcher in Jerusalem.[6] An undated act made at the time of Archbishop Atton of Arles (1115–1126), which mentions la Trinquetaille and Saint John the Baptist, but not the Holy Sepulcher in Jerusalem, may concern what was still an independent local foundation for the care of the sick, the poor, or pilgrims. Only in 1129, under a subsequent archbishop of Arles, Bernard Guérin (1129–1138), would its association with the Hospital of the Holy Sepulcher in Jerusalem be mentioned explicitly in a charter of donation to the hospital and its poor.[7]

Like many hospitals founded for local charitable purposes, for the care of the poor, pilgrims, lepers or for other reasons, la Trinquetaille became engulfed by the movement associated with the Hospital in Jerusalem and its support for the Crusades in the Levant. As was often the case, however, the charitable purposes of such hospitals in Western Europe were at odds with the needs of the Crusade, and it often appears that the provisioning of crusaders eventually outweighed the local needs which had spurred such foundations at the outset.[8] It is often difficult to know what donors expected when they gave land or rents to support the activities of this hospital, although sometimes it was burial.[9] The hospital also acted as the deathbed recipient of properties that were to be used as inheritance or dowry for the as yet unborn child of a donor's pregnant wife.[10] Elsewhere we see Bertrand de la Turre making a gift of claims to a large holding in return for being received as a *donatus*.[11] Only in a few cases would these charters involve what might be understood as service in the Holy Land. Thus, in 1194 William

[6] On such foundations at the time, see John H. Mundy, "Charity and Social Work in Toulouse, 1100–1250," *Traditio* 22 (1960), 203–87. On the gradual affiliation of such communities into an order, see parallel cases regarding the Cistercians in Constance H. Berman, *The Cistercian Evolution. The Invention of a Religious Order in Twelfth-Century Europe* (Philadelphia, 2000).

[7] *Cartulaire de la Trinquetaille*, nos. 257 and 258.

[8] On this, see Bronstein, *Hospitallers and the Holy Land*, pp. 80–1.

[9] *Cartulaire de la Trinquetaille*, no. 130 (1162), in which the donor gave himself for burial to the hospital of the poor.

[10] *Cartulaire de la Trinquetaille*, no. 231 (1201). Bernard Trabustul gave his own body for burial. Since his wife was pregnant, he also made arrangements for a posthumous son to be given in oblation to the hospital or a posthumous daughter to whatever religious community the hospital chose for her.

[11] *Cartulaire de la Trinquetaille*, no. 179 (1194) and no. 180 (1196). Received as a *donatus*, he bequeathed a place for pasture up to the saltmarsh, as well as cultivable lands with a boundary along the Rhône: "Loca ad pascendum usque ad stagna salata et terras laboraturas et fronteriam in Rodanum."

Porcelet made a gift of pasture rights near Saint Martin de la Crau for his brother, Porcellus, who had become a Hospitaller.[12]

This last donor, William Porcelet, and his brother Porcellus were part of a prominent local family in Arles, the Porcelet, whose association with new religious groups at the time must be seen as part of their assertion of power and piety.[13] Thus, in 1185 Rostagnus Porcelet had granted income from the rents over two mills for the fulfillment of his religious bequests to the hospital as executor of his estate.[14] Bertrand Porcelet gave "the entire honor" that he had at Venrella, including lands cultivated or fallow, wilderness, meadow, woods, marshes, and fisheries.[15] Similarly, in 1202 Adelaide Porcelet made a gift to the hospital of Saint Thomas of a piece of meadow at Bagnolet near other meadowlands that la Trinquetaille had bought from William Porcelet and Sacristana. Adelaide's gift was made so that she could be buried at the hospital of la Trinquetaille.[16] This Adelaide Porcelet was not the only woman giving herself for burial to the hospital of la Trinquetaille. Another apparently prominent donor was Agnes Pelliparia, who in 1210 gave herself and all her things to the hospital and to its preceptor, Arnold of Campagnoles, who accepted her gift and received her as a sister.[17]

Donors to both la Trinquetaille and Saint-Gilles included the Raymonds, counts of Toulouse with claims to Saint-Gilles, and the counts of Barcelona, who claimed Provence and eventually Montpellier.[18] As is

---

[12] *Cartulaire de la Trinquetaille*, no. 136 (1194): "Pro fratre meo Porcello qui hospitalarius effectus est … quoddam pascitum in Cravo quod vulgo cursorium dicunt."

[13] On this family, see Martin Aurell, *Une famille de la Noblesse Provençale au Moyen Age: Les Porcelet* (Avignon, 1986); and *Actes de la Famille Porcelet d'Arles (972–1320)*, ed. Martin Aurell (Paris, 2001).

[14] *Cartulaire de la Trinquetaille*, no. 63 (1185).

[15] *Cartulaire de Saint-Gilles*, no. 376 (1210), in which Bernard of Porcelet granted: "Totum honorem illum quem habebam vel habere debebam apud Venrellam vel in ejus toto tenemento sive territorio, in terris cultis vel incultis, sive in heremis, sive in pratis, sive in nemoribus, sive in paludibus, sive in taschis, sive in decimis, sive in gardiis vel in civitis vel in aquitis sive in piscationibus … continetur."

[16] *Cartulaire de la Trinquetaille*, no. 183 (1202). Sacristana seems to have supported the Cistercians instead. See Aurell, *Une famille*, pp. 162–5.

[17] *Cartulaire de la Trinquetaille*, no. 210 (1210). See also nos. 211 and 212.

[18] See *Cartulaire de Saint-Gilles*, no. 318 (1187), in which Raymond V count of Toulouse confirmed the rights of the community; and no. 344 (1190), where the count of Barcelona and Provence and king of Aragon gave pasture rights in the same region and not far from the house of Cistercian monks of Franquevaux. *Cartulaire de Saint Gilles*, nos. 260 (1114–50) and 186 (1182) were grants of free passage from the counts of Barcelona, who controlled this part of Provence. See also *Cartulaire de la Trinquetaille*, no. 206 (1150) for a donation from Raymond count of Barcelona.

clear from Cistercian records as well, religious communities usually acquired such grants, often involving exemption from tolls or taxes, from both parties disputing control, thus covering all of their bases. For the Raymonds or their rivals, however, it was a means of asserting their rights. These were lands only recently created by the river and its floods, as drainage of swamplands and the new construction of channels and levees caused a constant change in the landscape itself. This and the rich resources available in this region on the boundary between Languedoc and Provence, between the counts of Toulouse and the counts of Barcelona, meant that much of the territory was under dispute among civic authorities as well as among monastic and other religious corporations.[19]

Whether primarily providing funds for local charitable purposes or collecting monies and supplies to be send to the Levant, hospitals like la Trinquetaille became important owners of land, managing their holdings with an entrepreneurial spirit similar to that often ascribed to the new twelfth-century monastic orders, like the Cistercians. Many gifts and sales to such hospitals involved land, but some of the contracts concern rentals to tenants who contracted to bring land under cultivation or to ameliorate its condition by clearance and drainage or the planting of vineyards and trees to maintain the levees and drainage ditches.[20] The community might put out land on lease, such as one in which the tenant was to carry the Hospital's share of the produce from the fields up to a boat on the Rhône belonging to the monks.[21] Rights to fishing in channels, streams and fishponds were granted to the Hospital along with other rights to property, as in the conveyance to la Trinquetaille of rights "in the territory called Torremier from the Rhône river up to the Garrigue, whether in land, water, meadows, woods, fisheries, hunting rights or in whatever is there."[22]

La Trinquetaille's holdings in the Rhône delta make the community's documents particularly useful for the history of the ecology of

---

[19] Carraz, *L'Ordre du Temple*, pp. 108–21; and Berman, *Medieval Agriculture*, p. 82 on grants from rival authorities, and *passim* on monastic entrepreneurs. On exclusive zones, see Berman, *The Cistercian Evolution*, pp. 189–220.

[20] Aline Durand, *Les paysages médiévaux du Languedoc (XIe–XIIe siècles)* (Toulouse, 1998).

[21] *Cartulaire de la Trinquetaille*, no. 22 (1181).

[22] *Cartulaire de la Trinquetaille*, no. 145 (1193): "Quicquid quocumque iure quocumque modo habeamus vel habere debeamus in territorio quod apellatur Torum de Serico, a Rodano usque in Garrigam, sive sit in terra, sive in aqua, sive in pratis, sive in boschis, sive in piscationibus, sive in venationibus, sive in quocumque sit vel esse possit."

eel-fishing in the Middle Ages. This is explicit in the charters concerning access to fishponds or limits on eel trades, as well as in the lists of rents in eels owed the hospital, much of it for local consumption, but some possibly for shipment to crusader fronts as well. In addition to those witnessing the accumulation, consolidation and development of its properties, la Trinquetaille's charters include lists of annual payments made to the hospital. These are often rents in kind and not cash. Religious communities like la Trinquetaille received wood for heating and cooking (possibly for making barrels as well), salt for various needs, and cloth for clothing and bedding. Most often, however, these payments were for the foods necessary for their ascetic life-styles and to provide alms for the poor: payments in wheat or flour from a lord or lady's mills; quantities of wine produced by specific vineyards; or rights to annual deliveries of fish (salmon, herring or eels) which provided protein for the weekly or Lenten fasts.[23]

The la Trinquetaille cartulary contains several lists of annual rents payable to that hospital, two of which mention payments of eels. One is an early undated list of rents paid to la Trinquetaille by a single individual, Pons Fossi; it includes payment of fifty eels annually.[24] More noteworthy is a list from 1208, which contains three annual deliveries of eels to the hospital. Two of these were made by individuals identified

---

[23] See, for example, Paris, A. N. LL1595, Cartulaire de Saint-Antoine, fol. 36v–37r (1261), for a gift of fur intended for cloaks for the nuns. There were also rents designated for the purchase of food, such as the 30 livres granted annually by Louis IX in 1244 for the purchase of herrings at Arques. His mother Blanche of Castille also gave eight muids of wheat to the community at Pontoise. See *L'abbaye de Maubuisson, histoire et cartulaire, publiés d'après les documents inédits*, 2 vols. (Paris, 1890–1913), no. 112 (1244), and no. 300 (1239). For further examples, see Paris, B. N. Latin MS 13892, "Cartulaire du Lys," no. 39 (1311), in which Philip IV promised not to pay in cash but in kind; *La Grand Cartulaire de la Sauve Majeure*, ed. Charles Higounet and Arlette Higounet-Nadal, with Nicole de Pena (Bordeaux, 1996), no. 733 (n.d.), in which Peter of Albinion gave to la Sauve the church of Saint John of Sestiers, and a vivarium in front of that church, and no. 708 (1131), in which another donor gave the eighth part of the fish coming to market; *Cartulaire de l'abbaye de Voisins*, ed. Jules Doinel (Orléans, 1861), no. 121 (1238), in which the nuns received the jambage from the pig harvest; and *Cartulaire de Lézat*, ed. Paul Ourliac and Anne-Marie Magnou, 3 vols. (Paris, 1984), where references to donations of fish appear in well over twenty charters (see, for example, nos. 34, 75, 95, 147, 237, 259, 268, 333, 341, 373, 577, 608, 1052, 1100, 1101, 1102, 1225, 1329, 1536, 1571). For a discussion of the monastic diet at Cluny, which included the consumption of eels and many other kinds of fish, see Scott G. Bruce, *Silence and Sign Language in Medieval Monasticism: The Cluniac Tradition, c. 900–1200* (Cambridge, 2007), pp. 79–83.

[24] *Cartulaire de la Trinquetaille*, no. 262 (early twelfth century): "Poncius Fossi cum uxore sua et cum filiis suis, censum donat ad Ospitale in Fosso: VI sextarios de frumento, et alios VI vini, et V solidi, et V modios sale, et L anguillas, et unum porcum … et XII sextarios inter panem et vinem, et unum porcum."

with the castle or neighboring town of Fos-sur-Mer, and a third from a member of the Porcelet family:

> Beatrix Melleda de Foz, sister of the late Rufus of the New Chateau, ought to give to the Hospital of Saint Thomas in perpetuity the annual rent of two quarterions of the best eels in the first week of Lent. Peter Berenga of Foz gives annually to the hospital of Saint Thomas one quarterion of the best eels on the Feast of the Purification of the Blessed Mary (Candelmas). William Porcelet gives annually 4 quarterions of eels to the Hospital of Jerusalem, that is, two to the Hospital of Saint Thomas and two to the Hospital of Saint-Gilles, both in the first week of Lent.[25]

I have seen only two other rents of eels paid to religious houses in the region. One was to the monks of Valmagne, a Cistercian abbey located near the coastal lagoons just west of Montpellier.[26] The other was to the Cistercian monks of Silvacane, an abbey located east of the Rhône not far from Aix. This last charter, which describes a payment made from a pond or enclosure called "nadal," which I would construe as a nursery pond, suggests that eel-trapping in this region was in the process of becoming eel-farming.[27] It is published among the Porcelet charters by Martin Aurell, who suggests that it is the desire to maintain a constant supply of such eels that results in mid-thirteenth-century regulations limiting the catch in certain areas by outlawing certain types of fish traps and light-fishing.[28]

---

[25] *Cartulaire de la Trinquetaille*, no. 202 (October 1208): "Beatrix Melleda de Foz, soror quondam Rufi de Castro Novo, debet dare singulis annis in perpetuum pro amore Dei Hospitali S. Thome duos quartairons anguillarum captals in prima scilicet ebdomada kadragesime. Petrus Berengarius de Foz donat singulis annis Hospitali S. Thome unum cartairon de anguillis captals in purificatione beate Marie. Guillelmus Porcellet donat singulis annis quattuor quartairons de anguillis Hospitali Iherosolimitano; Hospitali scilicet S. Thome II quartairons et Hospitali S. Egidii alios II quartairons in quadragesime, videlicet in prima ebdomada." This charter is also calendared in the *Cartulaire de Porcelet*, no. 263 (October 1208).

[26] Montpellier, A. D. Hérault, 1 mi. 260–61, "Cartulaire de Valmagne," vol. 1, fol. 141v. (1177), with reference to a rent of 150 good eels (*anguilas bonas*). It also mentions a fisherman of the monks of that abbey as a witness.

[27] *Cartulaire de Porcelet*, no. 406 (1264), in which William Porcelet grants to the Cistercian monks at Silvacane 60 eels to be paid at Christmas: "annuatim in Nativitate Domini apud abbati Fos LX anguillas pontezas, bonas, pulcras, et sufficentes, quos predicto domino abbati et per eum dicto monasterio assignavit in quodam bordigol qui vulgariter dicitur Nadal, qui bordigol est in stagno quod dicitur de Corente.

[28] Aurell, *Une famille*, pp. 139–40; and *Chartes des Porcelet*, ed. Aurell, nos. 376 and 406. There was an agreement between Lord Bertrand Porcelet and Lord Rostaing of Fos-sur-Mer, co-lords of the castle of Fos, and the inhabitants of the villa of Saint-Geniès represented by its syndics of Saint-Geniès, which included clauses concerning eel-fishing in several of the great salt-lagoons near Fos, Porcelet, and Saint-Geniès. This agreement forbade traps for eels called *seguas* in the ponds of Corente and Martigues,

Such acts listing levies of eels suggest that in the twelfth- and thirteenth-century Camargue, and perhaps more generally in the Rhône delta (extending western beyond Montpellier and eastward to Marseilles), fishing and the farming of eels (or fattening of small eels, called eelings, in enclosed ponds) was possible, perhaps increasingly so given environmental changes. In this context, rents in eels were not exotic, but local produce, well-suited to the Lenten needs of monastic and other religious houses. Although it may well be that eels were being harvested from these waters from ancient times on, it is likely that land clearance, swamp drainage, tree and vine planting and other amelioration undertaken in the central Middle Ages caused changes in the Camargue that provided new habitats favorable to eels and suitable for eel-farming. If so, this environmental development, like the increase of flooding in such estuaries, has not been taken into account as a factor of the population increase in Western Europe that began around 1050 AD.[29]

The changes in the Mediterranean shoreline around the Rhône delta in this period, which probably explain the rise, for instance, of Montpellier as a major city where no Roman city had ever existed, provided new opportunities for agricultural and pastoral enterprises. As Aline Durand has shown in a recent study of the ecology of the Mediterranean coast slightly further to the west, river courses in this area were extremely unstable in these centuries, changing with drainage and flooding caused by clearance further upstream.[30] In this situation, she sees the establishment of mill dams and mills along rivers like the Hérault and the Orb as a way to prevent even worse flooding. The regulation of river flow may have been significant in creating the slow-moving streams that we so often associate with eels, for instance, in the English fenlands.[31]

---

and light-fishing in the pond of Corente. We can only guess at what the outlawed eel traps were like. Light-fishing involved stunning fish at night with flashes of lantern light.

[29] On the issue of such changes in modern estuaries, see Rodger Doyle, "Darkness on the Water: Development Takes its Toll on Estuaries," *Scientific American* (April 2007), p. 26. For the medieval Roussillon and Tet River valley, see Laure Verdon, *La terre et les hommes en Roussillon aux XIIe et XIIIe siècles: Structures seigneuriales, rente et société d'après les sources templières* (Aix, 2001); and Sylvie Caucanas, *Moulins et irrigation en Roussillon du IXe au XVe siècle* (Paris, 1995).

[30] Aline Durand, *Les paysages médiévaux du Languedoc.*

[31] Graham Swift, *Waterland* (London, 1983). I owe this reference to Colin Gordon. My thanks as well to discussions of eel fishing with Jennifer Sessions, Marshall Poe, and over lunch in Vancouver in the spring of 2008 with Richard Hoffmann.

The Camargue comprised multiple channels and lagoons of the Rhône delta in the vicinity of Arles, Saint-Gilles, les Saintes-Maries, Aigues-Mortes, and Fos-sur-Mer.[32] French historians Daniel le Blévec and Alain Venturin describe the region vividly in their edition of the Saint-Gilles cartulary:

> Dependent on its waters, because more than in most places nature contributed to create its landscape ... [t]he twelfth-century charters from these flat areas of the lower Rhône and from regions further to the west along the Mediterranean littoral show a region still in flux, poorly settled, with large marshy zones, dead river-branches and abandoned channels or meanders called *lones*. Such alluvial lowlands, where land and water were closely associated, could be fertile if one devoted time and effort ... These marshy expanses provided their owners with considerable resources for fishing, hunting, wood, watering, and feeding flocks.[33]

Access to one of the great European rivers allowed fish and eels to swim far upstream after spawning, but also gave them access to salt and freshwater marshes in the ever-changing river delta. The la Trinquetaille documents as well as those from Saint-Gilles bear witness to a medieval landscape of rough pine woods, marshes, great salt-ponds, like the Étang de Vaccarès and the Étang de Berre, drainage ditches, levees, saltworks, reclaimed and newly planted fields, vineyards and meadows, where rapid environmental change promoted habitat for a variety of fish, but eels in particular.[34]

This marshland environment bordering on the Mediterranean is reflected in descriptions of property in the charters from la Trinquetaille and Saint-Gilles. Saltworks along the Mediterranean shore, at places such as Fornales, were conveyed to la Trinquetaille as separate holdings.[35] They were distinct from the many salty ponds and salt marshes,

---

[32] Durand, *Les paysages médiévaux du Languedoc*, discusses settlement in the area immediately to the west of the delta.

[33] *Cartulaire de Saint-Gilles*, p. vii.

[34] *Cartulaire de Saint-Gilles*, no. 18 mentions an *asavena*, described by the editors as a "fish engine." This must be the mobile fish seine called *asagena* described by Richard C. Hoffmann, "Medieval Fishing," in *Working with Water in Medieval Europe: Technology and Resource Use*, ed. Paolo Squatriti (Leiden, 2000), pp. 331–93, at 359–63. The amelioration of such lands whether by natural forces or human effort is implied in charters such as *Cartulaire de Saint-Gilles*, no. 39 (1179): "Scilicet ipsas terras cum omnibus accessionibus et pertinentiis et cum omni melioramento quod ibi natura vel industria vel labore vestro provenerit ... totum sicut omnes terre suprascripte continentur a domo vestra usque ad Fossam et secus Fossam a Rodano usque ad terram Bessonorum ... usque ad Rodanum."

[35] *Cartulaire de la Trinquetaille*, no. 77 (1199–1200).

as is apparent from a conveyance of rights to land adjoining a certain *vallatum* called Paganum (probably a retaining wall, dike or levee) located in the salt-marsh (*in stagnis salatis*) below Mas Thibert, not far from the land called Pecia Rotunda.[36] Not far from this was property acquired by la Trinquetaille in the territory called Feudum which extended from the levee called Paganum and the Rhône up to la Crau and the salt marsh.[37] Land was also given above la Vernède and near that held from the hospital by Bertrand of Terre, adjoining the Hospital's land on the east and extending up to the salt-marsh and the land of John Scaecole.[38] Clearance of uncultivated land at Cepas Auca Ceca is recorded in a charter of conveyance between two laymen of a wilderness (*heremus*) to be brought under cultivation. The charter tells us that the *heremus* in question was to be uprooted and planted with half its produce paid to the original owner, and the other half to the tenant. After twelve years the land would revert to the original owner.[39]

Often what a charter describes is property defined only by rivers or channels or dikes, such as this charter from 1195: "At Près Gaudin, from Mas Thibert and the levee at Esfollat up to la Crau and the Rhône River, and from the levee called Paganum up to Pecia Rotunda and the tenement of la Vernède and to the old levee on the land that belonged to Peter Raymond of Terre and up to the Lone (river branch) of Torremier ... cultivated or fallow, woods, meadows, lands or water."[40] Such references to channels and dikes attest to flooding and other changes in the Rhône riverbed, which led to the creation of new islands and had consequences for other holdings. Thus a charter from 1146 describes: "that land which was purchased from my wife Sibilla, which is on the spit of land (*corrigia*) in front of the tower of Peter Lawrence," and

---

[36] *Cartulaire de la Trinquetaille*, no. 167 (1194): "Medietatem cuiusdam vallati quod Paganum apellatur et est subtus Mastiberto in stagnis salatis non longe a quadam terra que dicitur Pecia rotunda."

[37] *Cartulaire de la Trinquetaille*, no. 162 (1194).

[38] *Cartulaire de la Trinequetaille*, no. 178 (1196).

[39] *Cartulaire de la Trinquetaille*, no. 18 (1143).

[40] *Cartulaire de la Trinquetaille*, no. 152 (1195): "Quod apellatur pratum Galdini et quicquid habeo vel habere debeo a Mastiberto et a vallato quod hapellatur Esfollat usque in Cravum et usque in Rodanum et usque ad vallatum quod Paganum apellatur et usque ad Peciam Rotundam et usque ad tenementum Vernete et usque ad vallata vetera terrarum que fuerunt Petri Raimundi de Terre, et est sciendum quod infra isto terminos clauditur Lona de Taramars, quicuiqd autem habeo vel habereo debeo infra predictos terminos sive sit cultum, sive inculturm, sie sit in bosco, sive sit in pratis, sive in terris, sive in aquis sive possideam per me vel per alium." See also *Cartulaire de la Trinquetaille*, nos. 153 (1195), 154 (1195) and 159 (1197).

promises that if the Rhône floods the spit "we will get other land elsewhere to replace it."[41] Such islands are described in several charters. Some appear to have been held in shares by a number of different owners, possibly reflecting a co-ownership scheme from the start.[42] Land rescued from marsh, river, and sea was described as newly created by the forces of nature. One charter described "the island that has been created (*nata est*) in the area in front of Mas Thibert, with all its length and width, now or earlier, and all the augmentations to it or other islands in the Rhône made by man or accident, from the transverse levee at Mas Thibert up to the other bank with the Rhône flowing in between."[43] Such "new lands" often extended to the sea itself, as when a donor gave "whatever I hold or possess, cultivated or not, in the territory called Torremier on the sea and up to the channel of Daira."[44]

Fishing rights, both ponds and fisheries, while occasionally granted outright, were more often included in the lists of the appurtenances of a particular piece of property. On the Êtang of Berre at Marignane (the modern site of the Marseilles airport), for instance, someone donated land for one plow (*araire*) with four oxen, a meadow, vineyards, places to build houses, and a boat free of rent in the fishery (*pescaria*) of Bulmonis.[45] A few mills are mentioned as well, although there is no indication that the fisheries in question were associated with those mills or their mill-ponds. They were possibly near the rocky area of the Alpilles or near les Baux. Thus a charter from 1208 makes reference to a mill at Méjean and also refers to land at the great marsh (*palud*) adjoining the fisheries of Gajan.[46]

In the Saint-Gilles documents there are similar descriptions of such newly-cultivated land associated explicitly with local aquatic ecosystems. In 1171, Saint-Gilles received "a tenement, whether cultivated or

---

[41] *Cartulaire de la Trinquetaille*, no. 259 (1146): "amiseritis tantum spacii predictam terram teneatis et habeatis, totum in ipsa terra recuperetis."

[42] *Cartulaire de la Trinquetaille*, no. 247 (1204); no. 248 (1203), concerning an island for which the *cens* will increase if it gets bigger; and no. 300 (1202).

[43] *Cartulaire de la Trinquetaille*, no. 161 (1194): "Insulam que ante Mastibertum nata est cum longitudine et amplitudine quam modo habet vel in antea habere poterit, et cum omni augmento nec non et omnes alias insulas que hominum labore vel casu fient, vel hascentur in Rodano ex transverso a ripa Mastiberti usque in ripam aliam, inter quas ripas Rodanus currit."

[44] *Cartulaire de la Trinquetaille*, no. 142 (1170–1200): "Quicquid possideo in territorio quod dicitur Torium Saric et habeo loco usque ad mare in heremis et cultis." See also no. 143 (1167).

[45] *Cartulaire de la Trinquetaille*, no. 5 (1180).

[46] *Cartulaire de la Trinquetaille*, no. 302 (1208).

not, stretching on either side of the Rhône River and up to the sea."[47] In 1204 it added to its holdings two meadows with adjoining marshes and all additions possible from that great marsh.[48] The boundaries of these properties were often the new drainage ditches, dikes or levees, such as those between the levees that divided the holdings of Peter Folcra and Durand with two channels coming out of the pond and one coming out of the meander in the river (*lona*).[49] As at la Trinquetaille, such holdings were often rented out, like "the honor at Maresme up to Tela Longa by such an agreement that the tenants will give a quarter of the harvest from the vines and tithes and from the cultivated fields, the tasque, and tithes, and a thirteenth of the rabbits as well as rent in the clearances made at Maresme superior."[50] Sometimes the linear borders of a property are described with unusal vividness: "Over the lands and meadows and marshes enclosed at the lower end by the land of Barjac, following a right line up to the levee of Pignolo, with a similar straight line up to the cultivated lands of the nuns of Saint Cesarius of Arles and thence by a right line up to the Rhône river."[51] Rights held by the priory of Saint-Gilles thus mimicked those held by la Trinquetaille with occasionally more description, such as that of land at Faraldo and its appurtences, whether land or marshes, or tamarisk groves or meadow or pasture.[52]

Such charters underline the importance of fishing in general, possibly fishing and salting fish for transport elsewhere, along with eel production in the Camargue. The descriptions of land in the la Trinquetaille

[47] *Cartulaire de Saint-Gilles*, no. 53 (1171): "Scilicet … in toto illo tenemento quid cingitur ab utraque parte flumine Rodani et protenditure usque ad mare ubicumque sit et esse possit, sive sit cultum vel incultum, aut quicquid dici aut intelligi possit."

[48] *Cartulaire de Saint-Gilles*, no. 51 (1204).

[49] *Cartulaire de Saint-Gilles*, no. 11 (1164): "Illum vallatum qui est inter honorem Petri Folcra et honorem Duranti Aguiloni, cum duabus braceriis ex stagno et cum una braceria ex lona."

[50] *Cartulaire de Saint-Gilles*, no. 37 (1194): "Honorem qui est a Maresme usque ad Telam Longam, tali lege quod de vineis donent Hospitali quartonem et decimam et de honore laborivo tascham et decimam et de cirrogrillis (rabbits) accipiant terciam decimam partem pro gardia et in rumpitis que fecerint a Maresme superius habeant unum blatum grossum."

[51] *Cartulaire de Saint-Gilles*, no. 82 (1197): "Super terris et pratis et paludibus que continentur a capite inferiori terre de Barjaco, recta linea usque ad vallatum de Pignolo et a vallato de Pignolo similiter recta linea usque ad Expellariam Sancti Cesarii et ab Expellaria recta linea usque ad Rodanum."

[52] *Cartulaire de Saint-Gilles*, no. 105 (1182): "Quicquid jure habeo in Faraldo nec in pertinenciis suis que jure hereditario mihi preveniunt, sive sint terras vel paludes vel tamaricia aut prata aut pascua."

cartulary as well as in that for Saint-Gilles provide a different view of the much discussed rural development of the twelfth and thirteenth centuries in Western Europe. Estuaries seem to have experienced particularly severe environmental change, some to the profit of owners, others to their loss. In this and in other estuaries, however, the consequences of reclamation activities upstream had their effects in places like the Camargue where the activities of the Hospitallers and their patrons can be seen to have been much more than reeling in, or trapping, eels alone.

## *Bibliography*


*Unpublished Sources*

Montpellier, A. D. Hérault, 1 mi. 260–61, "Cartulaire de Valmagne."
Paris, A. N. LL1595, "Cartulaire de Saint-Antoine."
Paris, B. N. Latin MS 13892, "Cartulaire du Lys."

*Published Sources*

*Actes de la Famille Porcelet d'Arles (972–1320)*, ed. Martin Aurell (Paris, 2001).
*Cartulaire de l'abbaye de Voisins*, ed. Jules Doinel (Orléans, 1861).
*Cartulaire de Lézat*, ed. Paul Ourliac et Anne-Marie Magnou, 3 vols. (Paris, 1984).
*Cartulaire de la Trinquetaille*, ed. Paul-A Amargier (Gap, 1972).
*Cartulaire du Prieuré de Saint-Gilles de l'Hôpital de Saint-Jean de Jérusalem (1129–1210)*, ed. Daniel le Blévec and Alain Venturin (Paris, 1997).
*L'abbaye de Maubuisson, histoire et cartulaire, publiés d'après les documents inédits*, 2 vols. (Pontoise, 1890–1913).
*Le Grand Cartulaire de la Sauve Majeure*, ed. Charles Higounet and Arlette Higounet-Nadal, with Nicole de Pena (Bordeaux, 1996).

*Secondary Works*

Aurell, Martin. *Une famille de la Noblesse Provençale au Moyen Age: Les Porcelet* (Avignon, 1986).
Berman, Constance H. *Medieval Agriculture, the Southern-French Countryside, and the Early Cistercians: A Study of Forty-Three Monasteries* (Philadelphia, 1986).
——. *The Cistercian Evolution: The Invention of a Religious Order in Twelfth-Century Europe* (Philadelphia, 2000).
——. "Construction de moulins, défrichements et crues au onxième et douzième siècles dans le bassin de la Garonne," *Moulins de France* 73 (January 2008), 6–11.
Bronstein, Judith. *The Hospitallers and the Holy Land: Financing the Latin East, 1187–1274* (Woodbridge, 2005).
Bruce, Scott G. *Silence and Sign Language in Medieval Monasticism: The Cluniac Tradition, c. 900–1200* (Cambridge, 2007).
Carraz, Damien. *L'Ordre du Temple dans la basse Vallée du Rhône (1124–1312): Ordres militaries, croisades et société méridionales*, Collection d'histoire et d'archéologie médiévales 17 (Lyon, 2005).

Caucanas, Sylvie. *Moulins et irrigation en Roussillon du IXe au XVe siècle* (Paris, 1995).
Doyle, Rodger. "Darkness on the Water: Development Takes its Toll on Estuaries," *Scientific American* (April 2007).
Durand, Aline. *Les paysages médiévaux du Languedoc (XIe-XIIe siècles)* (Toulouse, 1998).
Hoffmann, Richard C. "Medieval Fishing," in *Working with Water in Medieval Europe: Technology and Resource Use*, ed. Paolo Squatriti (Leiden, 2000), pp. 331–93, at 359–63.
Jordan, William Chester. *Louis IX and the Challenge of the Crusade* (Princeton, 1979).
Mundy, John H. "Charity and Social Work in Toulouse, 1100–1250," *Traditio* 22 (1960): 203–87.
Selwood, Dominic. *Knights of the Cloister: Templars and Hospitallers in Central-Southern Occitania 1100–1300* (Woodbridge, 1999).
Swift, Graham. *Waterland* (London, 1983).
Verdon, Laure. *La terre et les hommes en Roussillon aux XIIe et XIIIe siècles: Structures seigneuriales, rente et société d'après les sources templières* (Aix, 2001).

# REACHING FOR A 'NATURAL AUTHORITY': THE RHÔNE IN EIGHTEENTH-CENTURY LYON

Pierre Claude Reynard

Historians have long probed the relationship between man and nature, and for several generations now, most have steered away from the shoals of physical or cultural determinism. Many, particularly in France, have been carried along by the current of "possibilism," recognising the primacy of social issues but acknowledging the weight of natural parameters through an extended chronological perspective. At the post-modern end of the twentieth century, new environmental movements challenged such certainties and generated more reciprocal formulations that emphasize the density of exchanges between a society and its environment. This paper probes an enlightened series of urban projects to illustrate this interdependency and, more precisely, to demonstrate the heuristic character of a great natural feature – the river Rhône in eighteenth-century Lyon, France's second largest city.[1]

Cities are social phenomena *par excellence* where the relations between natural conditions and human choices are often less apparent than in rural settings. They have only slowly claimed a place within the still growing field of environmental history. Not surprisingly, early advances in this direction focussed on issues of pollution, often of urban origin even if felt further afield, or on the emergence of crucial infrastructure networks originally mapped by historians of technology, public health, or communications. In turn, such themes explain the

---

[1] For this starting point, see the concluding remarks in *Du milieu à l'environnement: pratiques et représentations du rapport homme-nature depuis la Renaissance* ed. Marie-Claire Robic (Paris, 1992), p. 239. For reflections on the relations between the "possibilism" familiar to the Annales movement and the roots of environmental history, see Robert Delort and François Walter, *Histoire de l'environnement européen* (Paris, 2001); or, more succinctly, François Walter, "Une histoire de l'environnement, pour quoi faire?," in *Milieux naturels, espaces sociaux*, ed. Elisabeth Mornet, Franco Morenzi, and Danielle Milliod (Paris, 1997), pp. 33–42; and Georges Bertrand, "Pour une histoire écologique de la France rurale," in *Histoire de la France rurale*, ed. Georges Duby and Armand Wallon, 4 vols. (Paris, 1975), vol. 1, pp. 35–113.

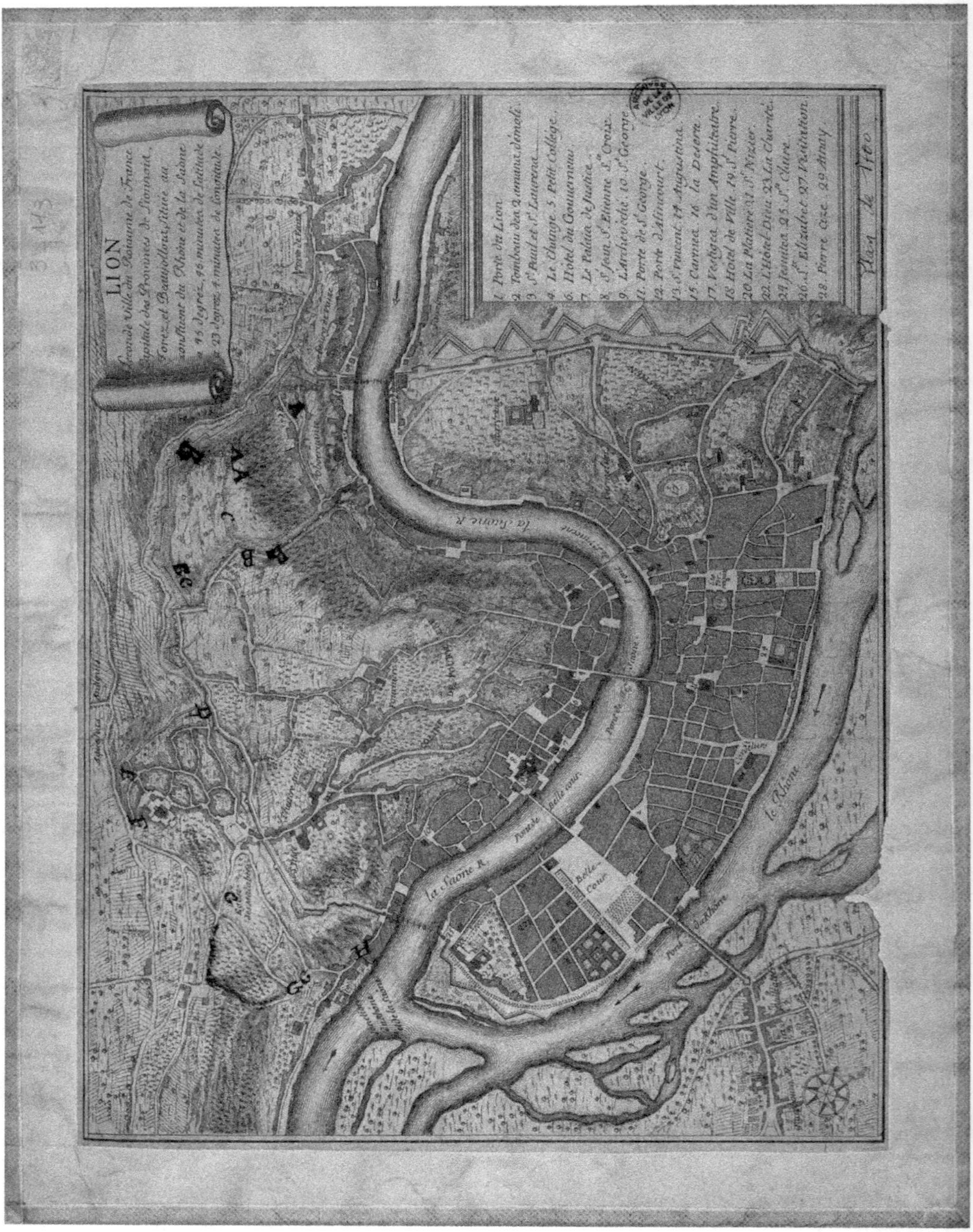
LION
la Saône R.
le Rhône
Belle-Cour

preponderance of modern studies, leaving earlier urban contexts in the shadows.[2] Yet, the very fact that cities accentuate the tensions between historical developments and their natural settings ought to draw our attention.[3] A first step is readily taken in this direction when the importance of the site upon which a city grew is acknowledged, but this recognition is often cursory or incantatory. Lyon's geography has received its share of comments in this regard, albeit mostly in general and timeless terms. The fate of the city has long been associated with its position on a major north-south European axis, a Lotharingian isthmus that is invoked to ground a range of influences, notably Italian, and a sustained commercial vocation. Even more regularly, Lyon's contrasted topography lends a tripartite structure to many arguments about the city: its steep hills to the west and north, a vast plain east of the Rhône, and the low-lying confluence of the Rhône and Saône separating them, are given distinct social and political personalities.[4] Yet, this triple polarity is young. It only became a reality during the second half of the nineteenth century, when the left bank of the Rhône was finally urbanised. More precisely, that far shore and the river that defines it had entered the city's imaginary a century earlier.

Until the eighteenth century, Lyon deliberately turned its back to the Rhône. From the 1730s onward, however, this great Alpine river shaped every urban project, successful or not (see Figures 1 and 2). In a few favourable decades, the city acquired an impressive eastern front along its right bank: a continuous embankment supporting a broad *quai*,

←

Figure 1. Map of Lyon, c. 1710. Copper engraving by Claude Brossette, published in *L'Histoire abrégée ou Éloge historique ou de la ville de Lyon* (Lyon, 1711).
Copyright Municipal archives of Lyon, France, 3 S 113.

---

2 A short survey of these patterns introduces the recent contributions gathered in *Resources of the City: Contributions to an Environmental History of Modern Europe*, ed. Dieter Schott, Bill Luckin, and Geneviève Massard-Guilbaud (Aldershot, 2005).

3 For a sharp illustration of what can be done in an early modern European context, see Elisabeth Crouzet-Pavan, *Venise: une invention de la ville* (Seyssel, 1997).

4 See, for example, comments by Fernand Braudel, *L'identité de la France* (Paris, 1986), 239–71, with the telling subtitles (Book One, chapter III and section I): "La géographie a-t-elle inventé la France?" and "Ne pas exagérer le rôle de l'« isthme français »." Modern presentations of Lyon distinguish the two hills of Fourvière and Croix-Rousse (the religious vocation of the first balancing the working-class history of the second), the decision-making Presqu'île, and the left bank that grew in the age of industry to accommodate a prestigious bourgeois quarter and many more working-class districts and work-sites. See Pierre-Yves Saunier, *L'esprit lyonnais – XIXe-XXe siècle* (Paris, 1995).

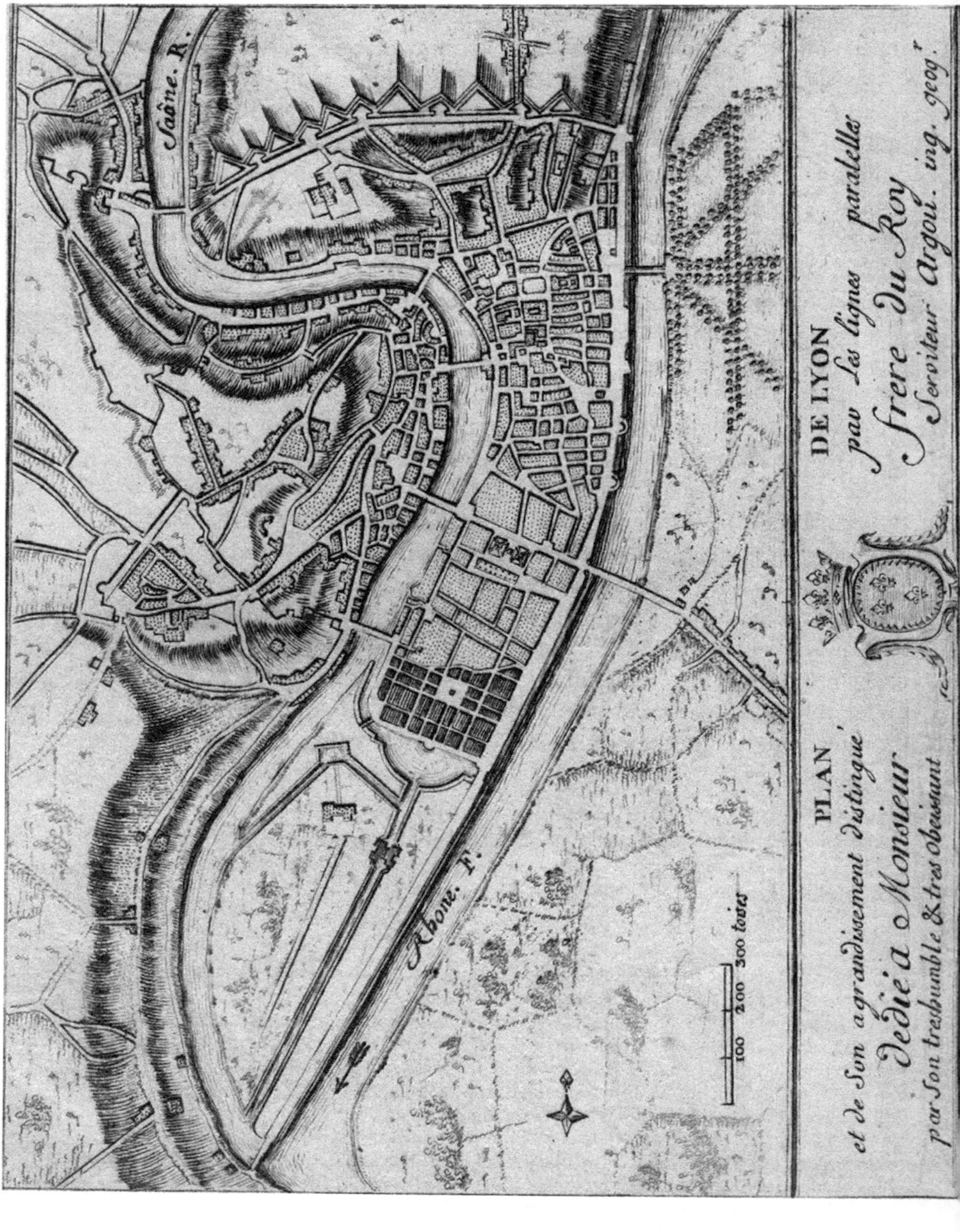
PLAN
DE LYON
et de son agrandissement distingué par les lignes paralelles
dedié a Monsieur frere du Roy
par son treshumble & tres obeissant Serviteur Argou. ing. geog.r
Saône. R.
Rhone. F.
100 200 300 toises

lined with tightly packed tall new buildings and, at its core, Soufflot's grand, yet appropriately linear, Hôtel-Dieu hospital. This "discovery of the Rhône" echoed long after a painful Revolutionary hiatus, and much of the urban thinking done in Lyon since has remained focussed on the Rhône, for good or bad. A little more than a century ago, an impressively landscaped Parc de la Tête d'Or tamed a great triangle of alluvial deposits to the immediate north-east of the city. After the Second World War, the Rhône's banks were sacrificed to automobile traffic and, to the south of the city, heavy industry. Not long after, a vast expanse that remained from the Rhône's past as a braided river, just upstream from the city, was preserved as a "natural area" – in the minimalist fashion of the moment.[5] Today again, while the problems facing the city evidently reach well beyond this river, its shores still agitate politicians and planners. Lyon is reclaiming the Rhône's left bank into a six kilometre long park to give an expansive sweep of the city's best face to strollers and fresh air to joggers. The north-eastern entrance of the city has also been re-engineered and re-landscaped around the river to provide a more natural and yet more stately approach. More ambitious, and less consensual, is yet another attempt to bring the confluence of Rhône and Saône within the urban core.[6] Because urban environments concentrate problems and solutions, they illustrate the extent to which nature provides both questions and answers, challenging common definitions of a city in the process.

The context of the great projects formulated in eighteenth-century Lyon is a classic combination of objective and subjective factors: substantial population growth and severe geographical constraints at a time when enlightened visions of healthier, more beautiful cities were

←

Figure 2. Map of Lyon, c. 1775. Engraving by Argout, geographer ("Plan de Lyon et de son agrandissement distingué par les lignes parallèles dédié à Monsieur frère du Roy par son très humble et très obéissant serviteur Argout, ing. Géographe." [vers 1775]).
Copyright Municipal archives of Lyon, France, 3 S 615.

[5] For a partial but first-hand account of the city's recent transformations, see Charles Delfante and Agnès Dally-Martin, *Cent ans d'urbanisme à Lyon* (Lyon, 1994).

[6] Current projects are placed in context in Jacques Béthemont and Jean Pelletier, "Lyon et ses fleuves: des berges perdues aux quais retrouvés," *Revue de géographie de Lyon* 65 (1990), 300–7; and in a municipal pamphlet: "Les rives des fleuves, le port et le parc," series "Les Notes," Lyon Confluence (Bibliothèque municipale de Lyon, 6900 E1 LYO). For a popular assessment, see "Lyon retrouve son fleuve," *Géo* (2003), 139–62.

spreading rapidly. Recent studies have confirmed the rising demography that was just grasped at the time: from fewer than 100,000 inhabitants at the beginning of the century to just under 150,000 on the eve of the Revolution. Existing suburbs could only accommodate a fraction of this increase. They were separated from the city by the Pierre-Scize narrows along the Saône to the north, by a lone and cramped stone bridge over the Rhône eastward, and by the long and appropriately named "chemin des Etroits" along the southern right bank of the Saône. Elsewhere, steep slopes inhibited expansion and slowed communications. The traditional attachment of religious orders and other wealthy interests to gardens further reduced available living space in Lyon to 150 hectares. If we also deduct from this paltry figure a meagre amount of public space, it is easy to understand why visitors always noted the impressive height of Lyon's mere 4,000 buildings, which sheltered both work and sleep. The bustle of immigrant accents in these dark streets and stairways only reinforced the sense of demographic pressure, further justifying the emergence of several plans to expand the city and bring the age's dreams of urban renewal within reach.[7]

Traditionally, Lyon's housing supply had grown through the reconstruction of existing buildings on a taller scale, although some dynamic periods did break new ground in their quest for space. The prosperity of the sixteenth century had, for instance, led to the development of peripheral lots within the walls, up until the Religious Wars. A century later, a new town hall "launched" the Terreaux neighbourhood to the north-east of the Presqu'île, and the vast Louis-le-Grand square drew part of the city's elite southward. Subsequently, several religious institutions replenished their coffers by selling land that was turned to housing.[8] These varied initiatives, like others elsewhere in France and in Europe, were intended to better house those who could afford it, celebrate existing powers, and enhance order and open space in the urban

---

[7] These issues are considered in detail by Maurice Garden, *Lyon et les Lyonnais au XVIIIe siècle* (Paris, 1970), Part One, ch. II; and Françoise Bayard, *Vivre à Lyon sous l'Ancien Régime* (Paris, 1997), pp. 105–11. Comparisons with other cities confirm the remarkable density of housing in Lyon: Jean Meyer, *Etudes sur les villes en Europe occidentale: milieu du XVIIe siècle à la veille de la Révolution française*, 2 vols. (Paris, 1983), vol. 1, pp. 135–6.

[8] The rate at which existing buildings were rebuilt has been estimated at 3.5% at mid-century (Garden, *Lyon et les Lyonnais*, p. 14). Françoise Bayard lists the lands ceded by religious orders (*Vivre à Lyon*, pp. 216–8).

fabric, if only tentatively.[9] In Lyon, none of these enterprises changed the shape of the city, until more ambitious proposals emerged by the middle of the eighteenth century.

* * *

Four large-scale operations thoroughly transformed the city after 1700. They were all driven by what the Rhône came to represent for the citizens of Lyon during the Enlightenment. The best known of these ventures is that of the "engineer" Perrache, the most lucrative raised the Saint-Clair district, and the most artistically outstanding was Soufflot's rebuilding of the Hôtel-Dieu. Complemented by a few more modest initiatives, such as the one carried out in front of the hôpital de la Charité, these undertakings thoroughly transformed the right bank of the Rhône. Even more instructive for our purpose was a fourth project initiated by the multi-talented Jean-Antoine Morand, designed to open the city to the left bank of the Rhône, that had hitherto remained pastoral and agricultural. The new neighbourhood would be known as the Brotteaux after the name of the sand and gravel banks characteristic of the braided channels of the pre-modern Rhône. This long and contentious sequence of projects exposes the central role played by the Rhône, an exceptional natural entity in this context, in conceptions of the city and its future at the dawn of the modern age.

Those who wished to see Lyon expand began to look seriously south to the confluence of the Rhône and the Saône at the end of the 1730s. The city had just purchased the Moignat island and the gardens of the abbey of Ainay at the foot of the ramparts that still guarded that southern end. A few memoranda praised the idea of expansion, but the Consulate remained too timid and financially pressed to act decisively. It merely leased the land under terms that called for plantations to consolidate these fragile deposits. This was a modest but also traditional approach: a powerful carrier of sediments, the Rhône regularly created

[9] These key themes are placed in context in Allan Braham, *The Architecture of the French Enlightenment* (Berkeley, 1980); Richard L. Cleary, *The Place Royale and Urban Design in the Ancien Régime* (Cambridge, 1999); Annick Pardailhé-Gallabrun, *The Birth of Intimacy: Privacy and Domestic Life in Early Modern Paris* (Oxford, 1991); and Jean-Louis Harouel, *L'embellissement des villes: L'urbanisme français au XVIIIe siècle* (Paris, 1993). For short surveys of the French context, see Roger Chartier, "La ville-chantier," in *Histoire de la France urbaine*, ed. Georges Duby, 5 vols. (Paris, 1981), vol. 3, pp. 109–56; and Jean Nagle, "La ville de l'absolutisme triomphant: De François I à Louis XV," in *Paris: Genèse d'un paysage*, ed. Louis Bergeron (Paris, 1989), pp. 93–145.

and destroyed hundreds of islands, fostering a remarkable complex of rules and practices intended to provide some stability where nature offered only change.[10] The first major achievement was that of the city's most prestigious institution, the Hôtel-Dieu. Starting in 1741, a long stretch of old hospital buildings directly fronting on the river were rebuilt on a new alignment, set back more than 25m atop a broad embankment that stretched from the ancient stone bridge north to the Terreaux. Thus was shaped Lyon's first true *quai* – the quai de Retz. This new avenue, further lit by the classic facade and elegant dome erected by the soon-to-be famous Jacques-Germain Soufflot (1713–1780), literally opened the city to the Rhône.[11] In turn, the impression of space and healthy freshness both created and revealed by the broad, fast-flowing, and cool river loudly stigmatised the noise and smells of the crowded old quarters, as well as the corruption of the smaller, busier, and sluggish river Saône on the other side of the Presqu'île. This contrast projected Lyon into a new era – calling for better conceived urban spaces and more pleasant surroundings.

Soufflot himself immediately realized the potential presented by a long gravel bed at the foot of the Croix-Rousse hill, just beyond the northern end of the new *quai*. Once connected to the Presqu'île and reclaimed, it served as a base for the new Saint-Clair district, which quickly became a sought-after address. Soufflot, his associates, and a score of investors who purchased lots on the reclaimed lands and built on them, made a fortune. The quai Saint-Clair extended the quai de Retz northward to open up a new road to the Bresse region, Geneva, and Central Europe, thereby sparing travellers and carters the taxing slopes of the Croix-Rousse. Most remarkably and somewhat surprisingly, since the buildings north of the hospital were independently

---

[10] These practices are the object of a current study by the author. See mentions in Jean-Paul Bravard, *Le Rhône, du Léman à Lyon* (Lyon, 1987), *passim*, and notably pp. 149–53 and 172–4; and A. M. Gurnell and G. E. Petts, "Island-Dominated Landscapes of Large Floodplain Rivers: A European Perspective," *Freshwater Biology* 47 (2002), 581–600. For these leases, see Archives municipales de Lyon (henceforth AML), BB 305, "Etude attentive du terrain de l'île Moignat ..." (1735) and lease to Joseph Gras (1740); purchase of the gardens of Ainay, DD 278, 16 Sept. 1738 (and several leases); plans by G. Delorme, 1738, DD 275; reports on the plans for expansion, DD 274 et 278.

[11] Among many studies of a celebrated career, one may turn to *Soufflot et l'architecture des Lumières* (Paris, 1980); and *L'oeuvre de Soufflot à Lyon: Etudes et documents* (Lyon, 1982).

designed, this dense and busy stretch of five or six-story facades cadenced by hundreds of tall windows eventually became an iconic image of the city.[12] This kind of pleasant architectural homogeneity (distinct from the more rigorous uniformity that framed a few exceptional settings, such as royal squares) was certainly rare, in Lyon and in early modern cities in general. In this case, we can only suggest that it was fostered by the new relationship with the Rhône, a process that can be articulated along at least three distinct perspectives. First, the new *quais* invited those who walked, rode, or drove by to appreciate these buildings as an ensemble. Second, the river itself provided a coherent, unifying landscape immediately across this new front. Third, the left bank, that was quickly becoming a favourite destination for Sunday strollers (see below), offered a broad perspective over these new constructions, a view that was framed by a hilly background.[13]

It was not long before attention turned again to the area south of the hospital, downstream from the Guillotière bridge. The idea of a southward expansion of the core peninsula was revived. The letters patent obtained by Antoine-Michel Perrache in 1770 authorized him to join the île Moignat to the Presqu'île by means of two levees along the right bank of the Rhône and the left bank of the Saône, pushing the confluence a mile further south. The former was to carry a road that would cross the Saône to join the highway south to the Languedoc. Where a branch of the river had separated the Presqu'île from the île Moignat, a canal would power the mills indispensable to the city and open an equally needed river port. On the lands thus gained, Perrache envisioned a mix of residential, industrial, and commercial buildings, with the open spaces and gracious vistas the age longed for.[14] After the successful opening of the "chaussée Perrache" which brought the entire right bank of the Rhône under control, this venture met with an endless series of setbacks. The bridge opened at the end of 1782 was washed

[12] Alain Charre and Catherine Servillat, "L'entreprise du quartier Saint-Clair," in *L'oeuvre de Soufflot*, pp. 21–6. The district was built around the 1760s (Archives départementales du Rhône, subsequently AD 69, 1 C 115, Correspondence de Lallié – Consulate, 26 Sept. 1775, 16 Dec. 1780).

[13] Discussions among three neighbouring owners trying to coordinate the elevation of their buildings may be found in AML, fonds Morand, 14 II 004, Correspondence Morand – Levet, dated 10, 19 Jan., 2, 7, 14, 16 Feb., 8 Mar., 1, 14 Aug., 18, 29 Sept., 9 and 29 Oct. 1760.

[14] Félix Rivet, *Le quartier Perrache (1766–1946): Etude d'histoire et de géographie urbaines* (Lyon, 1951). Contemporary commentaries in AML, 49 ii 3 (Fonds Fleurieu).

away a few weeks later by the Saône. Raising the lands newly connected to the Presqu'île to the required level took decades, leaving swampy depressions that obsessed an age terrified of stagnant waters. This unattractive district drew polluting industries at the expense of residential initiatives, and, late in the 1850s, a great railway station definitively cut it off from the core of the city. A little more than century later, a massive highway interchange again sealed off from the city this confluence area so painfully reclaimed! Characteristically, recurrent poor planning combined with geography – in this case, the inevitable isolation of a narrowing strip of land between two large rivers – to spoil an apparently sound idea.[15]

Nevertheless, by the last decade of the Old Regime, the right bank of the Rhône gave Lyon an open and monumental eastern facade almost a league long. The creation of long, vertical, and continuous stone embankments is characteristic of the age's desire for a controlled interface with rivers, even if, in many or most cities, these *quais* were only completed in the nineteenth century. This transformation is sometimes seen as marking the end of an era of conviviality, a positive term used to describe the bustle of pre-modern riverbanks.[16] It seems more fair to speak of the advent of another form of occupation of these unique sites, the promenade, that was just as popular but more contemplative and, thus, more likely to foster new urban visions. Nothing like this existed along the Saône, whose banks had long been densely used - a jumble of tight public and private spaces. For the first time, the citizens of Lyon could walk and ride at their leisure along the Rhône, a broad, underused, swift and steely river, and gaze at its flat, open east shore. The experience could not help but transform their ideas of the city.

The Hôtel-Dieu, which owned most of the land east of the Rhône and north of the Guillotière, did not fail to take advantage of this new awareness of the Rhône. In 1743, it was granted a royal monopoly for cable ferries. The hospital's Rectors went on to open up a wide "*grande allée*" and install the amusements expected by fair weather crowds. In the mid-1750s, they even considered building a bridge to enhance these

---

[15] Currently, blame for this latest mistake goes to the unhappy conjunction of local wishes to facilitate suburban access to the Presqu'île through a tunnel under the Fourvière hill and ministerial fiat appropriating that same tunnel for the freeway from Paris to the Mediterranean (Delfante and Dally-Martin, *Cent ans d'urbanisme à Lyon*).

[16] Jean Pelletier, "Sur les relations de la ville et des cours d'eau," *Revue de géographie de Lyon* 65 (1990), 233–9.

initial investments.[17] Early in the following decade, a still young architect *cum* entrepreneur invited this mighty institution to participate in the projection of the city across the river. At the time, the Hotel-Dieu was not in a position to take part in such an ambitious venture, even if it promised a windfall.[18] Undeterred, Jean-Antoine Morand (1727–1794) published his *Projet d'un plan général de la ville de Lyon et de son agrandissement en forme circulaire dans les terrains des Brotteaux* (1767). He proposed giving the city an almost perfectly circular shape, some three kilometres in diameter. The layout respected the natural (and partially fortified) limits formed by the hills and the former confluence (thereby excluding Perrache's rival project), while establishing a new district in a large semi-circle on the left bank. The two rivers would irrigate the new city in a majestic "Y." Across the Rhône, houses, shops, and warehouses would rise on an orthogonal plan aerated by several public squares. A second bridge was to link this new grid to the thriving Saint-Clair and Terreaux quarters.[19]

This vision was not without precedent. In 1733, the regional director of the rising department of civil engineering, the Ponts & Chaussées, had mused that the construction of dykes along the left bank of the Rhône would open up a "new continent" to which all incommodious trades could be exiled.[20] Nevertheless, despite such interest, Morand's project did not fare well. The city tentatively agreed to the building of a second bridge over the Rhône, on condition that it would not have to finance it. For its part, the Hôtel-Dieu firmly opposed all of Morand's undertakings for over a decade – in good part because such speculation sharply divided the elites who dominated this institution, the city, and

---

[17] Agreement of the Consulate, privilege for cable ferries, and other investments of the Hôtel-Dieu documented in AML, DD 308, fo. 125, 30 Aug. 1740, and in the Archives of the Hospices civils of Lyon, A 19, 28 Mar. 1743, and E 17, proceedings of 7 Dec. 1766. See also AML, 14 II 020, "Projet d'un pont sur le Rhône …," 1754, and "Essai ou projet sur les avantages … ."

[18] At a raucous meeting only months before Morand made his proposal, the hospital's Rectors had declared a moratorium on investments, on the left bank in particular, notably because of the great sums spent over the previous decades to rebuild its riverfront buildings in style (Proceedings of the Bureau de l'Hôtel-Dieu dated 31 Dec. 1766, Archives of the Hospices civils de Lyon, E 17).

[19] AML, "Plan circulaire" (plan and texts), 14 II 018 and 020; proceedings of the Consulate, BB 335 (20 May 1767). The map that accompanied this "Plan circulaire" is often chosen to illustrate eighteenth-century urban planning (see, for example, *Histoire de la France urbaine*, ed. Duby, vol. 3, p. 441).

[20] AD 69, 1 C 163, Report, Deville to the Consulate, 7 Mar. 1733. Other proposals in *Forma Urbis: Les plans généraux de Lyon du XVIe au XXe siècle* (Lyon, 1997).

its real-estate market. Nevertheless, Morand and his associates were able to build a toll bridge that proved very profitable. In 1780, Morand, who owned a substantial lot in the area, finally reached an agreement with the Hôtel-Dieu to subdivide the land for the future Brotteaux district. Sales were slow, however, and the Revolution, the siege of 1793, and the ensuing demographic shock stalled all development.[21] Early in 1794, Morand himself was guillotined for having participated in the defence of his bridge against revolutionary troops. It was only in the second half of the nineteenth century that Lyon really came to occupy the left bank, and this, tellingly, only after the Rhône had spoken again. The two exceptional floods of 1846 and 1856 made it clear that the development of the low-lying left bank would require more money, but also new thinking. Most notably, an effective system of protection from the river would have to combine a full embankment with what was finally becoming official policy – the protection of vast wet-lands above the city where flood waters could spill. This was not a new idea – the preservation of flood plains had already been given serious consideration in the seventeenth and eighteenth centuries along the Loire, albeit without sustained implementation. The destructive powers of the Rhône and the imperious needs of a great city combined to make it a reality (supplemented by further reforestation initiatives in the Alps).[22]

Long before that day, the Rhône structured all the debates spurred by Perrache's *confluent* project and Morand's Circular Plan. The eighteenth century decried the chaos of old "gothic" cities and its leading lights dreamed of new urban spaces.[23] Such reasoned enthusiasm, however, coexisted with a more organic vision, namely that a city lived off the concentration of its energies. In Lyon, the Rhône turned the minds of many towards the city's future, but hopes of rational growth

---

[21] Josette Barre and Paul Feuga, *Morand et les Brotteaux* (Lyon, 1998). Agreement between Morand and the Bureau de l'Hôtel-Dieu dated 24 Dec. 1780 (Archives of the Hospices civils de Lyon, A 20). The city of Lyon had been declared "counter-revolutionary" early in June 1793, taken by revolutionary troops three months later, and subjected to heavy repression.

[22] Jean-Paul Bravard sketches the key stages in the evolution of efforts to control the Rhône in *La Chautagne: Dynamique de l'environnement d'un pays savoyard* (Lyon, 1981), esp. p. 67 and following, with reference to the law of 25 May 1858 that protected flood plains upstream from cities. For early considerations, see Roger Dion, *Histoire des levées du Val de Loire* (Paris, 1961), p. 164 and following. See also Danielle Poinsart and Gilles Salvador, "Histoire de l'endiguement du Rhône à l'aval de Lyon (XIXe siècle)," in *Le fleuve et ses métamorphoses*, ed. François Piquet (Paris, 1993), pp. 299–313.

[23] These reflections are still best seen within their context in Jean-Claude Perrot, *Genèse d'une ville moderne: Caen au XVIIIe siècle* (Paris, 1975).

confronted fears of debilitating expansion. In many contexts, geography allowed for intermediate strategies, experiments on a lesser scale that would ease the transition to an outward-looking city. Often enough, for example, suburbs had grown and acquired specific functions that facilitated their integration into an urban centre.[24] Just as frequently, when crumbling city walls finally came down, the boulevards that replaced them emphasized the artificial nature of city limits, linking the inside and outside as much as separating them. The Rhône and its open left bank, on the contrary, heightened the contrast between the traditional forms of the city and the prospects of unlimited expansion. This confrontation stimulated the imagination of some but also alarmed others. The Rhône became an unavoidable participant in every discussion.

* * *

Among the initial objections to Morand's plan to expand the city eastward were those related to the many borders which the Rhône materialised in Lyon. The river marked the eastern limit of the city's excise, the Consulate's major source of revenue. A second bridge would increase the number of businesses clustered just beyond the gates; municipal finances would suffer, as would those of the elite burghers who enjoyed the right to sell wine from their estates free of tax within the city. Similarly, the owners of the eminently lucrative great royal tax farm, the Ferme Générale, worried that a new toll line further east, not backed by a large river, would be costly and more porous. The city would also have to purchase the rights to justice over the new district to provide its citizens with a coherent judicial framework.[25] Furthermore, the Rhône separated the provinces and *généralités* of the Lyonnais and Dauphiné, key administrative and fiscal divisions, and a new bridge would redraw the routes to Italy and central Europe. Little could therefore be done without the consent of Grenoble, seat of the Dauphiné's administration, and, naturally, Versailles itself.[26] Finally, if this re-centering of the city was to increase the number of ports and commercial spaces along the

[24] Jean-Pierre Poussou, "De la difficulté d'application des notions de faubourg et de banlieue à l'évolution de l'agglomération parisienne entre le milieu du XVIIIe et le milieu du XIXe siècle," *Histoire, Economie, Société* 15 (1996), 339–51.

[25] AML, 14 II 021, Trudaine to the intendant, 20 Mar. 1772; 14 II 004, Morand to Levet, 8 June 1779; and "Mémoire sur les inconvénients …," DD 308, 1788. Earlier, the city had purchased the justice rights to the Guillotière and the island Moignat (AML, BB 360, 23 and 24 Sept. 1705; DD 274, 29 Dec. 1740).

[26] Records of exchanges related to these issues in AD 69, 1 C 115.

Rhône, it also inevitably threatened established interests. For instance, the Guillotière, a rambling but lively string of inns and warehouses at the end of the only bridge across the Rhône for several hundred kilometres, stood to lose from the creation of a second crossing and a new neighbourhood on the left bank.[27]

The Rhône, a mighty river which would not be substantially "controlled" until the massive damming projects of the post Second World War era, also challenged Morand's projects in a number of technical ways.[28] The relatively narrow arches of the wooden bridge that he proposed to build would hinder navigation, particularly since they would stand scarcely more than a kilometre upstream from the existing stone bridge, itself an old structure with too many piers. This was a short distance for such a fast flowing river in the days before steam. Many residents of the modish Terreaux and Saint-Clair quarters also feared that the minimum height required for boats to clear the bridge's arches would be such that the resulting level of its western abutment would "bury" their houses. The Consulate, for its part, objected to the expenses necessary to connect the streets of the Presqu'île to the bridge and redesign drainage systems from the districts of the Terreaux and Saint-Clair. Typically, the matter could not be resolved by means of any "expert" or scientific measurements, since everything depended on the arbitrary choice of high water levels at which the barges would be required to cease their activities.[29]

Unlike many other rivers, the Rhône has traditionally divided more than united the regions through which it flows. Its sheer width and

---

[27] AML, 14 II 020, "Réponse des recteurs de l'Hôtel-Dieu," 7 Aug. 1769, and petition, 14 May 1772; BB 339, "Observations du Consulat," 19 Dec. 1771. AD 69, 1 C 115, 8 Jan. 1782, "Mémoire sur la nouvelle route … ." See also the statements made at the public inquiry for the registration of the letters patent given to Morand (AML, 14 II 020, December 1771). The Lyon bridge was the only one between Sault-Brénaz and Pont-Saint-Esprit, a distance of over 250 km.

[28] The Rhône occupied a central place in the modernisation of France. See Sara B. Pritchard, "Reconstructing the Rhône: The Cultural Politics of Nature and Nation in Contemporary France, 1945–1997," *French Historical Studies* 27 (2004), 765–99; and Gabrielle Hecht, *The Radiance of France: Nuclear Power and National Identity after World War II* (Cambridge, 1998).

[29] Among the numerous complaints by users of the river, see for example those received by Trudaine, the merchants' Provost, and by de Lallié (AML, 14 II 021, respectively 20 Mar., 18 Apr., and 18 May 1772). See also "Observations du sr. Morand …," 21 Dec. 1771, AML, DD 308. Several reports and letters reflect the residents's fears (AML, DD 308, 15 Apr. and 27 May 1772; 14 II 020, Morand to Guyot, 23 July 1768; Flachon – Morand, May–June 1772; Morand – Soufflot, March–May 1772; 14 II 021, de Lallié's report, 18 May 1772; and Morand's note, 28 Oct. 1774).

year-round volume, as well as its fast current, unstable banks, rapid changes of water level, and, above all, great sediment load, explain the dearth of bridges and, consequently, towns and cities spread to both banks. Many deemed Morand's bridge unworkable.[30] Others, more in touch with the ambitions and tastes of the time, argued that only a stone bridge would do. Although many wooden bridges were built to meet the considerable communication needs of the age, the Enlightenment could not help but prefer the purity of forms and the obvious solidity of stone to a perishable material such as wood. And, since an effective science of materials resistance was still decades away, all analyses were bedevilled by the increasing gap between the emerging professions of architects and engineers.[31] Discussions were particularly intense in this case because what might be sufficient on the Saône or the Seine might not resist the Rhône. Still, Morand found solid backing in Paris, and his bridge lasted over a century.[32] When we remember the notorious frailty of bridges in the early modern period (a problem well illustrated in Lyon by the accident-prone bridges over the tamer Saône), we are justified in thinking that the Rhône's reputation had forced all parties to take these challenges particularly seriously. This was only the fifth bridge across the Rhône south of Geneva, and if it was not particularly innovative, it was very well built. In this light, it is interesting to note that this same river gave rise, a few decades later, to a true innovation, that of the suspension bridge.[33]

The Rhône's record of disorder raised other objections. Armed with a list of famous floods reaching back to the year 592, Morand's opponents posited that in obstructing the riverbed, the bridge would force

---

[30] AML, 14 II 020, public inquiry (Dec. 1771); and Morand to Soufflot, 28 Feb. 1772. Abel Châtelain, "Les ponts du Rhône: Etude de géographie humaine," *Les Etudes Rhodaniennes* 19 (1944), 109–39; *La frontière: unir – diviser* (Givors, 1991); and Jean Frebault, Jean Dellus, and Martine Rivet, "Lyon ville fluviale," in *La ville et le fleuve* (Paris, 1989), pp. 37–46. Efforts to control the Rhône at Lyon are documented in AD 69, 1 C 162 and 163.

[31] Antoine Picon, *Architectes et ingénieurs au siècle des Lumières* (Marseilles, 1988); and *L'idée constructive en architecture*, ed. Xavier Malverti (Paris, 1987).

[32] These issues are well documented in AML, 14 II 020 and 021. Especially feared were winter ice breakups, more frequent in the early modern period than now (see for example AML, BB 252, 1694; 255, 1697; 270, 1709; 362, 1783; etc.). A first serious test came in January 1789. After several anxious days, it appeared that one of the most brutal ice break-ups in memory had destroyed only a few posts. Morand proudly recounted his relief in the *Journal de Lyon* dated 5 Feb. 1789.

[33] Michel Cotte, "Le premier grand pont à suspension de fil de fer construit par les frères Séguin à Tournon – Tain (1822–25)," in *La ville et le fleuve*, pp. 177–85.

the opening of a new channel further east, a potential disaster for the city and its recent right-bank investments. Just as serious, given the anxieties of the time, was the fear that more frequent flooding would turn the left bank into a marsh.[34] Somewhat contradictorily, the province's engineer, de Lallié, stated that preserving the Brotteaux as a spillage basin was vital to the safety of the existing city itself.[35] Yet, each objection was also a challenge eagerly taken up by some – the Age of Enlightenment had a taste for technical wagers which promised greater control over nature. The scientific and administrative institutions needed for a reasoned evaluation of new solutions were beginning to develop, but ideas that originated outside these young circles could still be heard, especially if the principle of royal privilege meant that they would not strain the treasury. And if the Rhône was less integrated into the daily lives of Lyon's citizens than the Saône, and more feared, it also fascinated many. Despite their longstanding "prejudice against the banks of the Rhône and its influence," in the words of Morand, crowds regularly gathered to survey this first new bridging of the river in centuries.[36] At least some of those who came must have sensed that this crossing of the Rhône would change the city.

* * *

Traffic problems were yet another topic of interest at the time. A second bridge would lighten congestion over the old one, facilitate maintenance of both, and permit a one-way system whenever circumstances dictated. Enlightened and optimistic individuals could also hope that it would force the opening of much-needed new east-west streets across the Presqu'île.[37] Some people were even willing to believe that pushing the confluence down-river would lower the Saône, allowing for the

---

[34] Comments by the Rectors of the l'Hôtel-Dieu and deliberations of the Consulate (AML,14 II 020, 7 Aug. 1769 and BB 339, 19 Dec. 1771). A counter-argument suggested that the multiplication of private properties would spread the cost of flood prevention measures, hitherto met by the city and its hospital (AML, 14 II 020 and DD 308, "Mémoire présenté aux Comtes de Lyon," April 1768).

[35] AML, 14 II 021, report by de Lallié, 18 May 1772.

[36] AML, 14 II 020, "Observations sur la dernière réponse de l'Hôtel-Dieu …," 14 Aug. 1769; and 14 II 021, Morand to Soufflot, 5 June 1774.

[37] While in Venice, Morand had observed that traffic flowed better along narrow, well defined lanes than open expanses (AML, 14 II 020, Report on the "Plan circulaire," 1766 and 1767; Morand also argued that his circular design minimized distances within the city). The crowds drawn to the launch of a hot-air balloon in 1784 called for a one-way traffic system (AML, 14 II 019).

opening and re-ordering of its banks (see below for references to the Chapuis project). This was a point of importance, because all those dreaming of urban improvement readily acknowledged the difficulties involved in renovating ancient quarters.[38] Once embraced, the Rhône, it seemed, would reach to the very heart of an old city.

The Rhône also offered opportunities to further the still limited but nonetheless real process of social and functional segregation in Lyon. It was quickly noted that a new bridge would spare the public the inconveniences of the ferries where the social elite was forced to rub shoulders with crowds that, even in their Sunday best, remained ill-behaved. Perhaps more substantively, the plains across the river could receive some institutions and businesses that needed to move away from residential areas to safer and cheaper ground. Noxious or dangerous professions had long been relegated to the outskirts of this and other cities, but a new bridge would allow taking full advantage of the remarkable protection that the Rhône could offer in this regard.[39] No less importantly, the Rhône's left bank held the promise of a mighty speculation that would replenish the coffers of the Hôtel-Dieu, an institution essential to social peace in this great manufacturing city. Lyon's fragile social climate would also benefit from new grounds for markets, fairs, warehouses, and stables that would augment and regulate supplies, and even, some keen entrepreneurs hoped, help to maintain salaries at "reasonable" levels. Moreover, in what may perhaps be termed an instance of poetic justice, the new bridge would also facilitate the daily removal of construction rubble and other less pleasant by-products of urban life.[40] The key point in all these matters was that the Rhône offered both a barrier and a link. Its width, quite exceptional by European standards, would shield the Presqu'île from many risks and problems. At the same

---

[38] AML, DD 278, "Third" and "Fourth and last" reports. Emmanuel Leroy Ladurie and Bernard Quillet, "Un urbanisme frôleur," in *Histoire de la France urbaine*, ed. Georges Duby, 5 vols. (Paris, 1981), vol. 3, pp. 439–81.

[39] Pierre Claude Reynard, "Public Order and Privilege: Eighteenth-Century French Roots of Environmental Regulation," *Technology and Culture* 43 (2002), 1–28. Fire was naturally a familiar concern in Lyon as elsewhere: François-Régis Cottin, "La protection contre l'incendie à Lyon sous l'Ancien Régime," *Bulletin de la Société historique, archéologique et littéraire de Lyon* 30 (2000), 139–64.

[40] These and other arguments were made repeatedly by Morand and his supporters (several drafts in AML, 14 II 020), as well as some of his predecessors ("Mémoire à Monseigneur le Contrôleur général ...," early 1740s, and "Premier mémoire," AML, DD 278). See also "Mémoire présenté aux Comtes de Lyon," April 1768 (AML, 14 II 020 et DD 308).

time, a second bridge would bring its left bank within easy access, granting the space indispensable to the kind of order the age wished to impose upon cities, and facilitating trade and communications with regions further afield.

The left bank would also meet the recreational needs of diverse social groups. The Rectors of the Hôtel-Dieu analysed the composition of the crowds hurrying to the Brotteaux, and generously concluded that "the common people being in need of amusements, it is fitting that they forget their sorrows in a whirl of pleasure [in order to] be all the more industrious." They planned to anticipate their expectations to better fill the coffers of their institution.[41] Morand also wanted an open plan and low density for the new district to rejuvenate the physical and mental strength of a weary city, and the bridge's very profitable toll accounts quickly demonstrated the tremendous appetite of the citizens of Lyon for recreation across the Rhône.[42] The river was more than a line or a neutral expanse separating old and new urban spaces, or even work from leisure. Its breadth and cool, swift waters offered an immediate break from the congestion and pollution of the old city and held the prospect of a wholesome alternative. The river both attracted people to, and prepared them for, a revitalizing experience. Even today, while both banks of the Rhône are fully built up, its many bridges offer a series of strangely removed breaches in the fabric of a dense city.

The integration of the Rhône into the city seemed to allow for, or even, demand a fundamental reworking of the existing urban order because the river rendered the city's anatomy more intelligible and thus more perfectible. We know what advances in medicine contributed to the early-modern understanding of cities, viewed increasingly as complex systems of organs, functions, and flows. The same

[41] AML, 14 II 020, "Projet d'un pont sur le Rhône ..." (1754) and "Essai ou projet sur les avantages ... Hôtel-Dieu ..." (ca. 1760). The left bank remained devoted to popular entertainment for many years: Puitspelu, "Distractions de jadis aux Brotteaux," *Rive Gauche* 154 (2000), 27–9; and Olivier Zeller, "En marge du privilège: petits spectacles et théâtre amateur à Lyon (1785–1787)," in *Théâtre et spectacles hier et aujourd'hui* (Paris, 1991), pp. 83–101.

[42] AML, 14 II 004, correspondence Morand – Levet praising the peacefulness of the Brotteaux, 2 July 1776, 31 May, 14, 27 June, and 19 Aug. 1777. Popular interest in the left bank was evident on the week-end the bridge opened, when more than a fifth of the city's population paid its toll: AML, 14 II 024, or Claude Mermet, "Les comptes du Pont Saint-Clair, 1771–1773," in *Hommage à Morand: A l'occasion du prêt à usage des papiers Morand de Jouffrey*, ed. Jeanne-Marie Dureau, Claude Mermet, and Marie-Félicie Perez (Lyon, 1994), pp. 95–107.

medical perspective also imposed a border, for a healthy body cannot be indeterminate. Many, along with Rousseau for example, spoke out against the dissoluteness of large urban areas and the lethargic weakness of gargantuan bodies.[43] Morand was well aware of the role of the Rhône in this regard. To meet this desire for closure, he repeatedly proposed to confine the Brotteaux within the double semi-circle of a wall and a canal. While adding to the defence of the city (and thus justifying his demand for a royal subsidy), these two features would also ease the work of customs officials, increase the number of ports, and drain off high water. This expensive project was never carried out. Nevertheless, the idea appears to have met initially with everyone's approval, a reminder of the importance of limits in modern conceptions of a city.[44]

* * *

The eighteenth century is generally viewed as a period of transition during which many towns and cities dismantled their walls and reached out to the open country beyond. In so doing, they were freeing themselves from both Latin and Germanic traditions which had dictated that the urban exception be visible and defended. It is true that the idea of a city as a unique and fragile set of political, social, and economic equilibria was being increasingly challenged through the early modern age by rising royal power, developing exchanges of all kinds, and urban growth itself. Nevertheless, up until the nineteenth century, even the most ambitious plans did not abandon the idea of controlling access to urban centres. The French monarchy had long neglected the fortifications around its capital, but it still sought to limit its growth and regulate even the most basic forms of traffic in and out of Paris – the "*mur des Fermiers généraux*," a customs wall, was one of its last great initiatives. Successive regimes continued to build "walls beyond the walls," right up until today's dreams of green belts. After all, even urban utopia are bounded.[45]

---

[43] Marcel Roncayolo, *Les grammaires d'une ville: Essai sur la genèse des structures urbaines à Marseille* (Paris, 1996), esp. p. 319; and Perrot, *Genèse d'une ville moderne*, p. 22.

[44] AML, 14 II 004, letters by Morand, May and June 1777.

[45] Relevant reflections in Chartier, "La ville-chantier," pp. 121–2; *La ville au XVIIIe siècle* (Aix-en-Provence, 1975); Christopher Friedrichs, *The Early Modern City 1450–1750* (London, 1995); Mona Ozouf, "Architecture et urbanisme: l'image de la ville chez Claude-Nicolas Ledoux," *Annales: Economies, société, civilisations* 21 (1966), 1273–1303; Antoine Picon, "De la composition urbaine au 'génie urbain': Les ingénieurs des Ponts et Chaussées et les villes françaises au XIXe siècle," in *Les langages de la ville*, ed.

Early modern suburbs were often a cause for official concern, being less policed and more likely to harbour unregulated practices and trades, legal or not. The Rhône only enhanced this familiar worry. The new bridge had barely been completed, when a serious accusation reached the minister responsible for Lyon, Bertin, suggesting that political opponents were meeting in secret at Morand's new house in the Brotteaux. The charge was critical because the city was in the throes of a political crisis in the wake of Laverdy's municipal reform and Turgot's subsequent coming to power – two sensitive episodes of what would eventually be seen as a pre-revolutionary age.[46] Many feared that a new district across the Rhône would prove particularly dangerous for Lyon, a city of plotting immigrants eager to escape beyond such a mighty screen. Such fears were still powerful enough during the riots of 1786 to trigger the cancellation of a modest deferment to 23 o'clock of the closing time of the city gates leading to the Brotteaux.[47] Others feared that rapid expansion of a new district could only be achieved at the expense of older neighbourhoods, thus threatening a delicate real estate balance that was, evidently, a powerful social and economic agent. Both Morand's and Perrache's plans were declared too extensive as soon as they became public.[48] Fears of unchecked growth, however, focussed on the former project, because the left bank of the Rhône was wide open to the East whereas the confluence was bounded by definition by two large rivers. A few years after the opening of the Saint-Clair bridge, when there were still only a handful of houses in the Brotteaux, rumours circulated that this new district would soon overtake the whole of the old city! In his pioneering demographic work, the *Nouvelles recherches sur la population de la France*, Messance felt it necessary to deplore the popular misconception that the number of empty houses in Lyon had grown since the end of the 1770s, that is, since the opening of a second bridge across the Rhône. As late as 1785, the Rectors of the Hôtel-Dieu argued that

---

Bernard Lamizet and Pascal Sanson (Marseille, 1997), pp. 169–77; and Marcel Roncayolo, *Lectures de villes: Formes et temps* (Marseille, 2002), ch. 3.

[46] AML, 14 II 004, Morand to Soufflot, 11 Sept. 1776, and Levet to Morand, 26 Apr. 1777. The Brotteaux notably attracted Masonic lodges. For the political context of Lyon, see Garden, *Lyon et les Lyonnais*; and Bayard, *Vivre à Lyon*.

[47] AML, DD 308, "Mémoire sur les inconvénients …," 1788; this authorization had been granted in 1781 to facilitate sales of land across the river (AD 69, 1 C 299, Poster, 24 Nov. 1781).

[48] See public inquiry (AML, 14 II 020, Dec. 1771), or letter from the intendant to Vergennes (AD 69, 1 C 298, 4 Mar. 1784). AML, BB 335, deliberations of the Consulate, 20 May 1767, and DD 275.

land in the new district "only [had a] value based on opinion," which they explained as an artificial value arising from "the taste of a few for a rural retreat." Even those who allowed that Lyon had to expand would have preferred what they called, a more "natural" form of growth. To them, the Rhône participated so clearly and evidently to the definition of the city that the urbanisation of its left bank seemed artificial.[49]

Such opposition to "forced" growth was less a symptom of early liberalism than a reflection of a growing regard for the "laws of nature," or, alternatively, the echo of older commitments to the mercantilist belief in the concentration of economic forces.[50] In 1788, a report condemning the pernicious effects of the Brotteaux spelled out that a city's general interest lay in the "concentration of its strengths." A similar assumption had prompted two well-intentioned canons of the great Saint-Jean cathedral to invite Morand to move back to the Presqu'île, just after he had settled his family in the Brotteaux district.[51] Morand followed their advice, and few people moved to the left bank indeed for many decades. Obviously, many Lyonnais believed that crossing the Rhône threatened the tight network of exchanges which made the urban experience so profitable. This fear was partly grounded in the traditional understanding of Lyon as an emporium, a hotbed of business whose energy depended on the constant renewal of ever-precarious contacts. The city's silk manufacture was known world-wide, but its relationship to its sphere of influence remained more one of dependence than domination. Its wealth was still measured by the degree of congestion of its ports, particularly during the four large annual fairs. This is the image conveyed by the famous and very busy scenographic map engraved by François Cléric in 1720, and this is the vision that was still powerful enough two generations later to counter the Chapuis project. Prompted by the promise of lower water levels following the removal of the confluence further south,

---

49 AML, 14 II 004, Morand to Levet, 13 May 1777; Messance, *Nouvelles recherches sur la population de la France* (Lyon, 1788), p. 68. AD 69, 1 C 299, Rectors to Terray, 15-5-1785. The concerns caused by peripheral expansion were the subject of several comments in Maurice Garden and Yves Lequin, *Construire la ville aux XVIIIe-XXe siècles* (Lyon, 1983).

50 See the argument made by Pierre Roncayolo in the case of Marseilles, *Les grammaires d'une ville*, pp. 332–334. More generally, see Pierre Dockès, *L'espace dans la pensée économique du XVIe au XVIIIe siècle* (Paris, 1969), particularly introduction and chapter I.

51 "Mémoire sur les inconvénients …," AML, DD 308, 1788; AML, 14 II 004, Levet to Morand, 8 June 1779. The same fear of a loosening of social ties underlined the concerns of the merchants of Caen regarding the width of new boulevards (Perrot, *Genèse d'une ville moderne*, p. 570).

a company of investors under that name wanted to straighten the left bank of the Saône along the centre of the Presqu'île, and develop the lands thus freed or reclaimed. This venture was blocked by the chorus of voices and interests that saw the chaotic commercial activity that took place in this old and cramped heart of the city as crucial to its prosperity.[52] The city's two great waterways represented powerful and intelligible internal structures, but they also were obvious links with the outside world. As a result, they lay at the heart of both enlightened projects for rationalizing space and traditional definitions of a commercial city.[53]

This duality speaks to the very nature of the river, which is both agent and object. This is, of course, always true, theoretically, of a river or any other natural feature, but it is most obvious, most legible in an urban context. More precisely, this proposition becomes thoroughly relevant when the river in question enters the urban fabric. To the people of Lyon, the Rhône obviously had always had a presence in the city, with several functions and representations attached to it. This role was largely unquestioned, obvious, or rather, "natural." First, it was a border, then, a minor trade artery, as well as a liminal space where urbane rules faded. In the space of a few decades in the eighteenth century, however, the Rhône assumed a more active, fuller role in the making of the city. It directly engaged the citizens of Lyon in a process of urban re-creation for which the Enlightenment supplied the key themes and tools. The Rhône actualised the modernising potential of the age in Lyon. The desires and misgivings, accomplishments and delays of those fifty years turned on the discovery of the full potential of this river as both a piece of nature and an object of artifice. A more ambitious analysis, encompassing the nineteenth century, would similarly place the Rhône at the heart of the city's reluctant, yet dynamic, transition from an ancient trading capital to a powerful industrial centre.[54]

* * *

[52] Michel Morineau, "Lyon l'italienne, Lyon la magnifique," *Annales: Economies, société, civilisations* 29 (1974), 1537–1550; and Olivier Zeller, "Enjeux d'urbanisme à Lyon en 1777: Propriétaires contre promoteurs," *Bulletin du Centre Pierre Léon d'histoire économique et sociale* 1 (1995), 3–15. See also the original presentation by Paul Henry, "La Saône, artère vitale de Lyon au début du 18e siècle, d'après la Vue Scénographique de Cléric en 1720," *Centre Presqu'île* 23 (1998), 15–7.

[53] For a Parisian perspective, see Isabelle Backouche, *La trace du fleuve: La Seine et Paris (1750–1850)* (Paris, 2000), p. 55.

[54] Pierre Cayez, *Metiers jacquard et hauts fourneaux aux origines de l'industrie lyonnaise* (Lyon, 1978).

The debates spurred by the prospect of a decisive expansion of Lyon across the Rhône naturally fit within the urban and environmental concepts of the eighteenth century, the Age of Enlightenment. Many of the arguments encountered here may be found in other contexts – other cities with different histories and unique natural features. To speak only of Ancien Régime France, several distinct sets of tensions are readily apparent. The relationship between Paris and the Seine was evidently in the hands of the monarchy, and thus often shaped by national, rather than simply, urban priorities. That of Bordeaux and its mighty estuary answered to a volatile but rich transatlantic vocation, while the expansion of Marseilles was marked by a fateful royal intervention in 1666 but also by its venerable relation to the Levant. In Nantes, another dynamic port, private initiative played a key role in the modernisation of the city, notably through the reclaiming of an island on the river Loire.[55] In Lyon, the Rhône initiated and dramatised all discussions. So long and so well a border, it rallied all the fears associated with the traditional definitions of the city. Yet, once "discovered," it also fostered just as many hopes for change, and led all parties to probe their understanding of the roots of the city's prosperity. To disregard the role of the Rhône in these debates is to ignore what linked all arguments, for and against expansion - the common ground that made all arguments possible.

Arguments for and against the expansion of Lyon beyond the Rhône can first be grouped around specific interests: professional, financial, political, and those representative of key eighteenth-century ideas that were meant to create beautiful, healthy, and orderly cities. We can then distinguish two contrasting analyses of the vitality and prosperity of the city. Those who wanted to urbanise the left bank applied an analysis that was functional and rational, presenting a perspective that would, eventually, sustain the new science of urban planning. The Rhône underwrote all (or at least most) of their plans. Their opponents often seemed to adopt a more defensive position, one that protected existing interests, doubted the promises of engineers, or refused to be convinced by the merits of rationalizing space. These trepidations, however, also had a positive basis. For those who feared crossing the Rhône, a city, and especially this one, lived off the confluence of energies it attracted. For them, the Rhône was there to signal, but also to enhance, this

---

[55] Guy Saupin, *Les villes en France à l'époque moderne, XVIe-XVIIIe siècles* (Paris, 2002) provides a survey of the evolution of important French cities, as well as further bibliographic references.

concentration, the very essence of a commercial hub. The functionalism of one camp conflicted with the essentialist convictions of the other, but the Rhône was at the heart of both perspectives. Those who opposed the expansion of Lyon across the Rhône saw it as a natural limit that was essential to the city's vitality. Conversely, their rivals argued that this river offered unique opportunities to modernise the city. Common to both perspectives is a wish to enhance their arguments through references to an indisputable power: the mighty Rhône. In both cases, this quest for a "natural authority" answered concerns raised by modernity's growing separation of the human race from its natural environment – in effect reaching back to the older, normative meaning of the word nature in terms appropriate to the age.[56]

This link may be exposed through a brief glance at the distinct choices made by two sisterly disciplines that often address the same issues. Perhaps because they are uniquely attentive to the physical complexity of different settings, geographers have freed themselves more readily than historians from the temptation of natural determinism. Urban geography, in particular, has long been human geography. To its practitioners, no city is defined by its natural setting, however remarkable.[57] Historians remain more reluctant to part with the concept of a set of fundamental natural data that may not determine the outcome of trends and events, but can never be ignored or circumvented with impunity. So keenly aware of the multiplicity of social possibilities, historians remain tempted by the recourse to a reference point which is stable because it is not human. That is, in effect, the contribution of the "possibilist" current dear to twentieth-century French (and other) historians, and that is why this approach can legitimately be understood as a step toward an environmental history. More at ease with the modern emancipation of humans from their world than historians, geographers respect it readily. Less scientifically trained, historians feel the need to transgress it more often.

The Rhône occupied a central place in the debates spurred by the need for more space in Lyon in the eighteenth century, and it has continued

---

[56] Augustin Berque, *Etre humain sur la terre: Principes d'éthique de l'écoumène* (Paris, 1996), particularly Part One: "Les bornes de la modernité"; and Kenneth Olwig, *Nature's Ideological Landscape: A Literary and Geographic Perspective on its Development and Preservation on Denmark's Jutland Heath* (London, 1984).

[57] Jean Pelletier speaks of the quasi absolute absence of "site determinism" ("Sur les relations de la ville").

to do so because such a key natural feature speaks of permanence in an uncertain world. It promised and continues to promise a solid foundation for every proposal. Evidently, this quest for a firm natural reference is but one dimension of our relationship to nature. Nevertheless, it deserves the attention of historians and, more particularly, of environmental historians interested in urban phenomena. The natural features of an urban site may or may not define what is possible or impossible at a particular time. More importantly, however, and well before such limits are tested, references to nature and, more precisely, to unique natural features, regularly speak to the hopes and fears of a society.

## *Bibliography*


Backouche, Isabelle. *La trace du fleuve: La Seine et Paris (1750–1850)* (Paris, 2000).

Barre, Josette, and Paul Feuga, *Morand et les Brotteaux* (Lyon, 1998).

Bayard, Françoise. *Vivre à Lyon sous l'Ancien Régime* (Paris, 1997).

Berque, Augustin. *Etre humain sur la terre: Principes d'éthique de l'écoumène* (Paris, 1996).

Bertrand, Georges. "Pour une histoire écologique de la France rurale," in *Histoire de la France rurale*, ed. Georges Duby and Armand Wallon, 4 vols. (Paris, 1975), vol. 1, pp. 35–113.

Béthemont, Jacques, and Jean Pelletier, "Lyon et ses fleuves: des berges perdues aux quais retrouvés," *Revue de géographie de Lyon* 65 (1990), 300–7.

Braham, Allan. *The Architecture of the French Enlightenment* (Berkeley, 1980)

Braudel, Fernand. *L'identité de la France* (Paris, 1986).

Bravard, Jean-Paul, *La Chautagne: Dynamique de l'environnement d'un pays savoyard* (Lyon, 1981).

——. *Le Rhône, du Léman à Lyon* (Lyon, 1987).

Cayez, Pierre. *Metiers jacquard et hauts fourneaux aux origines de l'industrie lyonnaise* (Lyon, 1978).

Chartier, Roger. "La ville-chantier," in *Histoire de la France urbaine*, ed. Georges Duby, 5 vols. (Paris, 1981), vol. 3, pp. 109–56.

Châtelain, Abel. "Les ponts du Rhône: Etude de géographie humaine," *Les Etudes Rhodaniennes* 19 (1944), 109–39.

Cleary, Richard L. *The Place Royale and Urban Design in the Ancien Régime* (Cambridge, 1999).

Cotte, Michel. "Le premier grand pont à suspension de fil de fer construit par les frères Séguin à Tournon - Tain (1822–25)", in *La ville et le fleuve* (Paris, 1989), pp. 177–85.

Cottin, François-Régis. "La protection contre l'incendie à Lyon sous l'Ancien Régime," *Bulletin de la Société historique, archéologique et littéraire de Lyon* 30 (2000), 139–64.

Crouzet-Pavan, Elisabeth. *Venise: une invention de la ville* (Seyssel, 1997).

Delfante, Charles, and Agnès Dally-Martin, *Cent ans d'urbanisme à Lyon* (Lyon, 1994).

Delort, Robert and François Walter, *Histoire de l'environnement européen* (Paris, 2001).

Dion, Roger. *Histoire des levées du Val de Loire* (Paris, 1961).

Dockès, Pierre. *L'espace dans la pensée économique du XVIe au XVIIIe siècle* (Paris, 1969).

*Forma Urbis: Les plans généraux de Lyon du XVIe au XXe siècle* (Lyon, 1997).

Frebault, Jean, Jean Dellus, and Martine Rivet, "Lyon ville fluviale," in *La ville et le fleuve* (Paris, 1989), pp. 37–46.
Friedrichs, Christopher. *The Early Modern City 1450–1750* (London, 1995).
*La frontière: unir - diviser* (Givors, 1991).
Garden, Maurice. *Lyon et les Lyonnais au XVIIIe siècle* (Paris, 1970).
Garden, Maurice, and Yves Lequin, *Construire la ville aux XVIIIe-XXe siècles* (Lyon, 1983).
Gurnell, A. M. and G. E. Petts, "Island-Dominated Landscapes of Large Floodplain Rivers: A European Perspective," *Freshwater Biology* 47 (2002), 581–600.
Harouel, Jean-Louis. *L'embellissement des villes: L'urbanisme français au XVIIIe siècle* (Paris, 1993).
Hecht, Gabrielle. *The Radiance of France: Nuclear Power and National Identity after World War II* (Cambridge, 1998).
Henry, Paul. "La Saône, artère vitale de Lyon au début du 18e siècle, d'après la Vue Scénographique de Cléric en 1720," *Centre Presqu'île* 23 (1998), 15–7.
*L'idée constructive en architecture*, ed. Xavier Malverti (Paris, 1987).
Ladurie, Emmanuel Leroy, and Bernard Quillet, "Un urbanisme frôleur," in *Histoire de la France urbaine*, ed. Georges Duby, 5 vols. (Paris, 1981), vol. 3, pp. 439–81.
"Lyon retrouve son fleuve," *Géo* (2003), 139–62 (no author cited).
Mermet, Claude. "Les comptes du Pont Saint-Clair, 1771–1773," in *Hommage à Morand: A l'occasion du prêt à usage des papiers Morand de Jouffrey*, ed. Jeanne-Marie Dureau, Claude Mermet, and Marie-Félicie Perez (Lyon, 1994), pp. 95–107.
Messance, *Nouvelles recherches sur la population de la France* (Lyon, 1788).
Meyer, Jean. *Etudes sur les villes en Europe occidentale: milieu du XVIIe siècle à la veille de la Révolution française*, 2 vols. (Paris, 1983).
*Du milieu à l'environnement: pratiques et représentations du rapport homme-nature depuis la Renaissance* ed. Marie-Claire Robic (Paris, 1992).
Morineau, Michel. "Lyon l'italienne, Lyon la magnifique," *Annales: Economies, société, civilisations* 29 (1974), 1537–50.
Nagle, Jean. "La ville de l'absolutisme triomphant: De François I à Louis XV," in *Paris: Genèse d'un paysage*, ed. Louis Bergeron (Paris, 1989), pp. 93–145.
*L'oeuvre de Soufflot à Lyon: Etudes et documents* (Lyon, 1982).
Olwig, Kenneth. *Nature's Ideological Landscape: A Literary and Geographic Perspective on its Development and Preservation on Denmark's Jutland Heath* (London, 1984).
Ozouf, Mona. "Architecture et urbanisme: l'image de la ville chez Claude-Nicolas Ledoux," *Annales: Economies, société, civilisations* 21 (1966), 1273–303.
Pardailhé-Gallabrun, Annick. *The Birth of Intimacy: Privacy and Domestic Life in Early Modern Paris* (Oxford, 1991).
Pelletier, Jean. "Sur les relations de la ville et des cours d'eau," *Revue de géographie de Lyon* 65 (1990), 233–9.
Perrot, Jean-Claude. *Genèse d'une ville moderne: Caen au XVIIIe siècle* (Paris, 1975).
Picon, Antoine. *Architectes et ingénieurs au siècle des Lumières* (Marseilles, 1988).
——. "De la composition urbaine au 'génie urbain': Les ingénieurs des Ponts et Chaussées et les villes françaises au XIXe siècle," in *Les langages de la ville*, ed. Bernard Lamizet and Pascal Sanson (Marseille, 1997), pp. 169–77
Poinsart, Danielle, and Gilles Salvador, "Histoire de l'endiguement du Rhône à l'aval de Lyon (XIXe siècle)," in *Le fleuve et ses métamorphoses*, ed. François Piquet (Paris, 1993), pp. 299–313.
Poussou, Jean-Pierre. "De la difficulté d'application des notions de faubourg et de banlieue à l'évolution de l'agglomération parisienne entre le milieu du XVIIIe et le milieu du XIXe siècle," *Histoire, Economie, Société* 15 (1996), 339–51.
Pritchard, Sara B. "Reconstructing the Rhône: The Cultural Politics of Nature and Nation in Contemporary France, 1945–1997," *French Historical Studies* 27 (2004), 765–99.

Puitspelu (pseudonym of Clair Tisseur), "Distractions de jadis aux Brotteaux," *Rive Gauche* 154 (2000), 27–9.
*Resources of the City: Contributions to an Environmental History of Modern Europe*, ed. Dieter Schott, Bill Luckin, and Geneviève Massard-Guilbaud (Aldershot, 2005).
Reynard, Pierre Claude. "Public Order and Privilege: Eighteenth-Century French Roots of Environmental Regulation," *Technology and Culture* 43 (2002), 1–28.
"Les rives des fleuves, le port et le parc," series "Les Notes," Lyon Confluence (Bibliothèque municipale de Lyon, 6900 E1 LYO).
Rivet, Félix. *Le quartier Perrache (1766–1946): Etude d'histoire et de géographie urbaines* (Lyon, 1951).
Roncayolo, Marcel. *Les grammaires d'une ville: Essai sur la genèse des structures urbaines à Marseille* (Paris, 1996).
——. *Lectures de villes: Formes et temps* (Marseille, 2002).
Saunier, Pierre-Yves. *L'esprit lyonnais - XIXe-XXe siècle* (Paris, 1995).
Saupin, Guy. *Les villes en France à l'époque moderne, XVIe-XVIIIe siècles* (Paris, 2002).
*Soufflot et l'architecture des Lumières* (Paris, 1980).
*La ville au XVIIIe siècle* (Aix-en-Provence, 1975)
Walter, François. "Une histoire de l'environnement, pour quoi faire?", in *Milieux naturels, espaces sociaux*, ed. Elisabeth Mornet, Franco Morenzi, and Danielle Milliod (Paris, 1997), pp. 33–42.
Zeller, Olivier. "En marge du privilège: petits spectacles et théâtre amateur à Lyon (1785–1787)," in *Théâtre et spectacles hier et aujourd'hui* (Paris, 1991), pp. 83–101.
——. "Enjeux d'urbanisme à Lyon en 1777: Propriétaires contre promoteurs," *Bulletin du Centre Pierre Léon d'histoire économique et sociale* 1 (1995), 3–15.

# DEFINING 'NATURAL' FISH COMMUNITIES FOR FISHERY MANAGEMENT PURPOSES: BIOLOGICAL, HISTORICAL, AND ARCHAEOLOGICAL APPROACHES

Wim Van Neer and Anton Ervynck

## *Introduction*

Most large river basins in Europe are heavily affected by severe human disturbances in the past. There is now a growing awareness, however, that this situation must change and that a durable use of water for present and future generations will only be achieved if the problem is addressed internationally. In an attempt to stimulate the improvement and the maintenance of the water quality of European waters, the so-called Water Framework Directive (WFD) was implemented in December 2000 by the European Union. This directive forms the basis for an integrated river basin management in the European Union and it requires that the Member States integrate the directive into their national legislations. The WFD stipulates that by the end of 2015 all European surface waters and groundwaters need to be of 'good quality,' meaning that the deterioration of these waters must be prevented and that the necessary measures need to be taken to protect, improve or restore their quality. The WFD further requires that the ecological quality of waters is monitored through a network of monitoring sites and that the results of this survey are reported in 2009, at the latest. It is, in addition, stipulated that through water management plans the ecological quality at these monitoring points should reach a 'good status' by 2015.

The ecological quality of waters – which depends on a number of factors such as the physico-chemical quality, the hydrological regime and the morphological characteristics of the basins – can be evaluated through the analysis of indicator groups, such as phytoplankton, aquatic vegetation, macro-invertebrates and fish fauna.[1] Among the quality

[1] Mats Wallin et al., *Final Guidance on Establishing Reference Conditions and Ecological Status Class Boundaries for Inland Surface Waters: EU Common Implementation Strategy (CIS) for the Water Framework Directive* (Luxemburg, 2003).

elements retained by the WFD for the classification of rivers, three parameters are listed with regard to fish, namely species composition, abundance and age structure.[2] To reach the goals set by the WFD, national and regional authorities are developing for various surface waters a 'fish index' that quantifies the deviation of the present ichthyofauna from the one that can be expected in an undisturbed, 'natural' condition (the so-called 'reference condition'). On a European scale, a more general fish index has been developed that should allow a comparison of the ecological status of all European waters.[3] Of course, the use of fish to monitor and evaluate the quality of ecosystems requires background knowledge of the structure of undisturbed or minimally affected environments. A major drawback, however, is that such pristine or high status conditions against which levels of human impact, such as overfishing or habitat degradation, could be measured are virtually nonexistent nowadays. Therefore, as an approach to establish riverine reference conditions, the WFD recommended the use of historical and palaeo-ecological data.[4]

The present paper will discuss the possible contribution of historical and archaeozoological data to the establishment of reference conditions. First, the methodology used by fishery biologists to evaluate and monitor water quality will be presented in general terms. Then a case study will show the fishery research carried out through time in the Scheldt, a major river basin in Belgium. Lastly, this paper will describe the biological information that is relevant for the compliance with the WFD and discuss the added value of historical and archaeozoological data from Flanders for an integrated approach to the evaluation of the quality of river basins.

## *Reference Conditions and Fish Indices*

In order to assess current levels of human impact, biological baselines need to be developed that ideally correspond to pristine or 'minimally

---

[2] Roger Owen et al., *Definition and Establishment of Reference Conditions: Water Framework Directive, REFCOND, Ispra* (Aberdeen, 2001).

[3] Alcibiades N. Economou, *Development, Evaluation and Implementation of a Standardised Fish-Based Assessment Method for the Ecological Status of European Rivers: A Contribution to the Water Framework Directive (FAME), Defining Reference Conditions (D3), Final Report* (2002); see http://fame.boku.ac.at/downloads/D3_reference_conditions.pdf

[4] Owen et al., *Definition and Establishment of Reference Conditions.*

disturbed' conditions.[5] Specific reference conditions need to be defined for each surface water body type and, in addition, class boundaries need to be set for the description of the quality of the waters evaluated (high, good, moderate, poor or bad status). When fish are used as the indicator group, through field surveys of the present-day waters that need to be assessed, the present composition of the freshwater fish fauna can be described in detail with information on the species present, their abundance, the lengths, weights and condition of the species, and so on. This information can then be compared to a reference condition that, ideally, should be the 'original,' undisturbed situation. In reality, such reference sites with pristine conditions are rarely preserved, which necessitates taking into account historical and palaeo-ecological elements that can help to model the original condition. The description of this former, minimally disturbed or pristine condition is less exact since it depends on the availability of data that decrease as one goes farther back in time. Moreover, varying sources of information yield data of different quality and nature.

On the basis of the differences found between the present ichthyofauna and the one defined in the reference condition, an assessment can be made of the present-day water quality, and decisions can be formulated regarding the target conditions that will be aimed for in the future. When defining the target conditions, options will very often be dictated by socio-economic reality and it will therefore be unrealistic to aim for a complete return to pristine conditions. Reduction of pollution, prevention of overfishing, and the re-introduction and restocking of fish can help to ameliorate water quality, but a restoration of, for instance, the original hydromorphology of the basins that would allow flooding is less realistic in economic terms. Similarly, it is unlikely that policy makers will agree to modify again the structure of the water bodies that have been adapted over the centuries for hydropower and shipping, except perhaps on a small scale. In general, there is a tendency to define target conditions that are realistic and therefore different from the pristine situation which is less desirable as an instrument for policy makers since it is not attainable. Therefore, fishery authorities tend to define target conditions that clearly represent a less disturbed past situation that is not too distant from the present-day conditions.

---

[5] Owen et al., *Definition and Establishment of Reference Conditions*; and Wallin et al., *Final Guidance on Establishing Reference Conditions.*

This, however, does not mean that the establishment of reference conditions is useless. The reference conditions indicate the framework against which the target conditions have to be defined. In a way, the target conditions are a step situated part way between the present conditions and the reference conditions.

As a tool to measure the condition of surface waters on the basis of fish as an indicator group, various indices have been developed since 1981 based on the assumption that there is a predictable relationship between the structure of a fish community and the physical, chemical and biological condition of a river system. The Index of Biotic Integrity (IBI) is a complex multi-metric index that was first developed for use in small warm water streams in central Illinois and Indiana.[6] The original version had twelve metrics that reflected richness and composition of the fish species, number and abundance of indicator species, trophic organization and function, reproductive behavior, fish abundance, and condition of individual fish.[7] The various parameters and the boundary classes can be adapted to the regional circumstances (from geological, morphological and biogeographical points of view) of the basin or to the particular, smaller water course under consideration.

## *The Scheldt Basin and its Fish Indices*

Following the methodology of Karr, IBIs have been devised for the two major river basins in Belgium, that of the Meuse and that of the Scheldt.[8] In this paper we concentrate on the river Scheldt (Dutch: Schelde;

---

[6] James R. Karr, "Assessment of Biotic Integrity Using Fish Communities," *Fisheries (Bethesda)* 6 (1981), 21–7.

[7] Environmental Protection Agency (EPA), *Metrics and the Index of Biotic Integrity* (2006); see http://www.epa.gov/bioindicators/html/ibi-metrics.html

[8] Karr, "Assessment of Biotic Integrity Using Fish Communities"; Jacques Didier, *Indice Biotique d'Intégrité Piscicole (IBIP) pour évaluer la qualité écologique des écosystèmes lotiques* (Namur, 1997); Patrick Kestemont et al., "Selecting Ichthyological Metrics to Assess River Quality Basin Ecological Quality," *Archiv für Hydrobiologie Supplement* 121.3–4 (1999), 321–48; Delphine Goffaux et al., *A Biotic Index of Fish Integrity (IBIP) to Evaluate the Ecological Quality of Lotic Ecosystems: Application to the Meuse River Basin, Final Report (User Guide Life 97 ENV/B/000419)* (Groenendaal, 2001); Claude Belpaire et al., "An Index of Biotic Integrity Characterizing Fish Populations and the Ecological Quality of Flandrian Water Bodies," *Hydrobiologia* 434 (2000), 17–33; Jan Breine et al., "A Fish-Based Index of Biotic Integrity for Upstream Brooks in Flanders (Belgium)," *Hydrobiologia* 522 (2004), 133–48; and Jan Breine et al., "A Fish-Based Assessment Tool for the Ecological Quality of the Brackish Schelde Estuary in Flanders (Belgium)," *Hydrobiologia* 575 (2007), 141–59.

F: Escaut), which is the largest basin on Flemish territory.[9] The Scheldt is a 350 km long river that finds its origin in northern France at Gouy near Saint-Quentin and empties into the North Sea near Vlissingen (Figure 1). The Scheldt enters Belgium near Tournai and runs northeast to Ghent, where it receives its main tributary, the Lys (Dutch: Leie; F: Lys). From its source to Ghent the river is called 'Bovenschelde' (Upper Scheldt). Downstream of Ghent, the river turns east, and near Antwerp it flows in a westward direction into the Netherlands. From Ghent to the Belgian-Dutch border the river is called 'Zeeschelde' and from the frontier to the river mouth it is designated as 'Westerschelde.' The modern topography of the lower reaches of the river is, however, different from the one in former, post-glacial times when the major discharge into the North Sea happened much further north through another branch, the 'Oosterschelde.' It was only in the eleventh or twelfth century AD that the 'Westerschelde' became the major outlet. Most data dealt with in the present overview are from the late medieval period and it can be assumed that the course of the Scheldt for this period was more or less comparable to the present one. Although the river has maintained its main course, since the late Middle Ages important changes have influenced its hydrology. Dikes have been built and lands have been embanked, reducing the lateral extent of floodwaters.[10]

Although there is a salinity gradient, the Scheldt can be considered as brackish between Antwerp and the Belgian-Dutch border, and as freshwater upstream of Antwerp. Salinity zones are relatively stable nowadays, despite seasonal and annual fluctuations.[11] The river can be subdivided following Huet's typology, which refers to aquatic zones defined by particular parameters such as stream velocity, river bed slope and cross-section, water temperature, oxygen content, substrate and vegetation.[12] The typology also defines the typical and auxiliary

---

[9] The analyses of the Flemish waters are carried out by the Flemish 'Research Institute for Nature and Forest' (INBO), a scientific institute of the Flemish Government in Belgium.

[10] Mark Van Strydonk and Guy De Mulder, *De Schelde. Verhaal van een rivier* (Leuven, 2000); and Alexander Van Braeckel et al., *Historische analyse van de Zeeschelde en haar getijgebonden zijrivieren, 19e eeuw tot heden: Report INBO.R.2006.29* (Brussel, 2006).

[11] Karline Soetaert and Peter M. J. Herman, "Estimating Residence Times in the Westerschelde (The Netherlands) Using a Box Model with Fixed Dispersion Coefficients," in *Major Biological Processes in European Tidal Estuaries*, ed. Carl H. R. Heip and Peter M. J. Herman (London, 1995), pp. 215–24.

[12] Marcel Huet, "Profiles and Biology of Western European Streams as Related to Fish Management," *Transactions of the American Fisheries Society* 88.3 (1959), 155–63.

Figure 1. Map of the Scheldt Basin.

fish species that can be encountered in each of these zones. Besides the estuarine zone, which typically harbors euryhaline fish (both marine and freshwater species that thrive in brackish water), the major part of the Scheldt basin in Flanders belongs to the bream and barbel zones.[13] More upstream waters belonging to the grayling and trout zones exist as well, but they are of limited extent in Flanders.[14]

IBI's have been developed for the estuary, for the bream and barbel zones and for the upstream zones of the river Scheldt.[15] Typically, the methodology involved surveys of hundreds of sites, using mainly electrofishing.[16] Subsequently, from a large number of candidate metrics a

[13] *sensu* Marcel Huet, "Profiles and Biology of Western European Streams."

[14] Jan Breine et al., "A Fish-Based Index of Biotic Integrity for Upstream Brooks in Flanders (Belgium)."

[15] Claude Belpaire et al., "An Index of Biotic Integrity Characterizing Fish Populations"; Jan Breine et al., "A Fish-Based Index of Biotic Integrity for Upstream Brooks in Flanders (Belgium)"; and idem, "A Fish-Based Assessment Tool for the Ecological Quality of the Brackish Schelde Estuary in Flanders (Belgium)."

[16] Electrofishing uses electricity to stun fish before they are caught. When performed correctly, the method results in no permanent harm to fish, which quickly return to their natural state.

restricted amount was selected using ecological criteria and statistical analyses. After scoring the metrics, they were then averaged to define the IBI as well as 5 integrity classes (from high status to bad). Once the IBI was tested, it could be applied to individual sites within the basin. Reference conditions have recently been defined for Flemish waters by water body type (for example, 'very large river,' 'large river,' 'small river,' 'brook,' and so on) that list the species that should be present and the abundance in which they should occur.[17] The approach adopted by the INBO, as developed by Vrielynck and colleagues, is summarized in Figure 2.[18]

### *Historical Data Used Thus Far for the Scheldt*

While defining the reference conditions for the Scheldt basin, the fishery biologists of INBO took data into account that they classified as being 'historical.' Those comprised mainly old fishery data from 1840 onwards, derived from scientific biological papers or books or from publications compiled within the context of attempts to promote and ameliorate freshwater fishing.[19] The work of Vrielynck and his colleagues can be seen as a continuation of earlier publications that recognised the potential of old fishery data for reconstructing the former fish fauna.[20] They are all mainly based on early, more or less scientific, reports about the Scheldt ichthyofauna, dating from 1842 onwards, and

---

[17] Ilse Simoens et al., *Monitoringsproject visfauna: Afleiden en beschrijven van systeemeigen referentieomstandigheden en/of maximaal ecologisch potentieel voor visgemeenschappen in elk Vlaams oppervlaktewaterlichaamtype, vanuit de - overeenkomstig de Kaderrichtlijn Water - ontwikkelde beoordelingssystemen op basis van vismonitoring (Onderzoeksopdracht nr.: VMM.AMO.SCALDIT.VISII)* (Groenendaal, 2005).

[18] INBO: Flemish 'Research Institute for Nature and Forest'; and Sven Vrielynck et al., *De visbestanden in Vlaanderen anno 1840–1950: Een historische schets van de referentietoestand van onze waterlopen aan de hand van de visstand, ingevoerd in een databank en vergeleken met de actuele toestand. (Rapport Instituut voor Bosbouw en Wildbeheer IBW.Wb.V.R.2002.89)* (Hoeilaart, 2003).

[19] Vrielynck et al., *De visbestanden in Vlaanderen anno 1840–1950.*

[20] Vrielynck et al., *De visbestanden in Vlaanderen anno 1840–1950*; Paul A. Van Damme et al., "Fish Species in the Lower Zeeschelde (Belgium): A Comparison with Historical Checklists," *Belgian Journal of Zoology* 124 (1994), 93–103; Joachim Maes et al., "The Composition of the Fish and Crustacean Community of the Zeeschelde Estuary (Belgium)," *Belgian Journal of Zoology* 127 (1997), 47–55; and Alain Vandelannoote et al., *Atlas van de Vlaamse beek- en riviervissen* (Wijnegem, 1998).

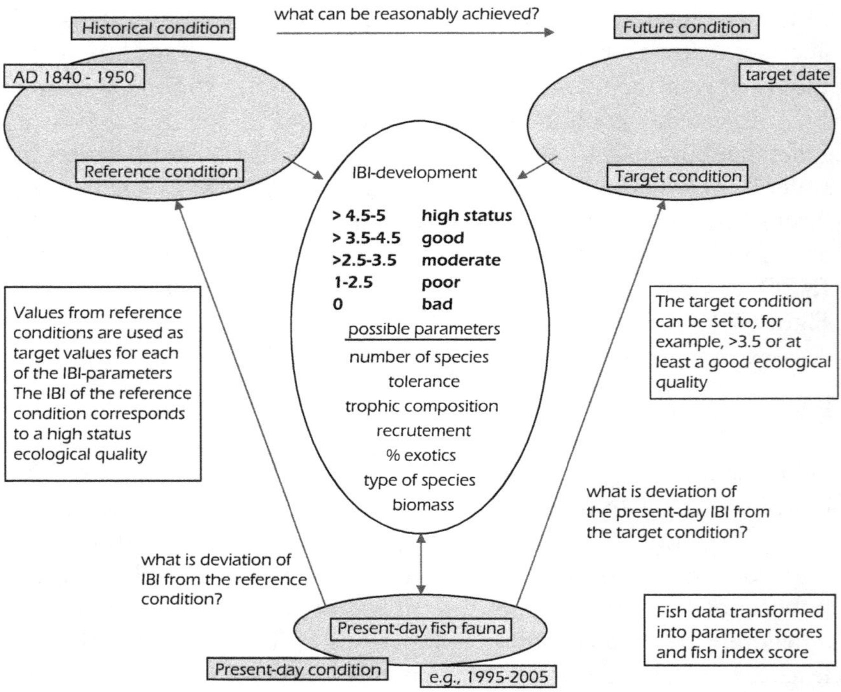


Figure 2. Relationships between the Fish Index and the Historical, Present-Day and Future Condition of Flemish Surface Waters (modified after Vrielynck et al., 2003).

on later publications by Lameere, Maes, Raveret-Wattel, Van Aelbroeck and Rentiers, Rousseau and colleagues, and Poll.[21] For the period from 1890 to 1943, the study by Vrielynck and his colleagues also relied heavily on the information provided by the Belgian fisheries journal *Pêche*

[21] Edmond de Selys-Longchamps, "Classe IV, Poissons d'eau douce," in *Faune belge 1: Indication méthodique des mammifères, oiseaux, reptiles et poissons observés jusqu'ici en Belgique* (Liège, 1842); idem, *La pêche fluviale en Belgique*, Extrait de l'Académie royale de Belgique, 2ième série, tome 22, n. 12 (Bruxelles, 1867); idem, *Révisions des poissons d'eau douce de la faune Belge*, Extrait de l'Académie royale de Belgique, 3me série, tome 14, n. 12 (Bruxelles: Académie royale de Belgique, 1887); Auguste Lameere, *Manuel de la faune de Belgique 1: Animaux non insectes* (Bruxelles, 1895); L. Maes, *Notes sur la pêche fluviale et maritime en Belgique* (Bruxelles, 1898); Casimir Raveret-Wattel, *Atlas de poche des poissons d'eau douce de la France, de la Suisse Romande et de la Belgique: avec leur description, moeurs et organisation*, Bibliothéque de poche du naturaliste 11 (Paris, 1900); Florent Van Aelbroeck and Emile Rentiers, *Poissons des eaux douces*

*et Pisciculture*, although the data available in that journal have not yet been fully explored.[22]

The data obtained from the aforementioned sources do not include all the parameters used when calculating an IBI from samples obtained during present-day fish surveys and this prompted the development of a fish index especially devised for historical data. The 'historical' fish index only includes number of species, number of non-indigenous species, the mean tolerance value and the proportion of predators. On the basis of these reference data, the deplorable condition of the present Flemish fish faunas could be demonstrated. For the twentieth century there appeared to be a general decrease in ecological quality from 'good' to 'critical' in many water courses, although this trend has been reversed recently in several waters.[23] This decrease in IBI during the last century has mainly been caused by a general decline in migrating species, an increase of introduced allochthonous species, an increase in the number of tolerant species, a decline in the number of sensitive rheophylic species and a decrease in the relative abundance of predator species.[24] Consequently, actions have already been taken to re-introduction species believed to be autochthonous, while poorly represented species have been artificially bred and restocked by INBO. Of course, such actions can only be undertaken when the overall water quality is already improving. Efforts have concentrated thus far mainly on endangered species that have little commercial value, such as brown trout (*Salmo trutta fario*), burbot (*Lota lota*), dace (*Leuciscus leuciscus*), pike (*Esox lucius*), chub (*Leuciscus cephalus*), tench (*Tinca tinca*) and bullhead (*Cottus gobio*).[25] The restocking and re-introduction policy tries to take into account as much as possible data from population genetic research in an attempt to preserve genetic diversity.[26]

---

*et saumâtres dans leur habitat* (Bruxelles, 1913); Ernest Rousseau et al., *La pêche fluviale en Belgique: Manuel à l'usage des pêcheurs et des gardes* (Bruxelles, 1915); Max Poll, "Contribution à la connaissance de la faune ichthyologique du Bas-Escaut," *Bulletin du Musée Royal d'Histoire Naturelle de Belgique* 21 (1945), 1–32; and idem, *Poissons marins* (Brussels, 1947).

22 Vrielynck et al., *De visbestanden in Vlaanderen anno 1840–1950*.

23 Vrielynck et al., *De visbestanden in Vlaanderen anno 1840–1950*, Table 20.

24 Vrielynck et al., *De visbestanden in Vlaanderen anno 1840–1950*.

25 Claude Belpaire, personal communication.

26 See, for example, Alain Dillen et al., *Onderzoek naar de biologie van de kwabaal (Lota lota L.), ter voorbereiding van het herstel van de soort in het Vlaamse Gewest (Rapport van het Instituut voor Natuurbehoud 2005.04)* (Brussel, 2005).

This first effort to use historical fishery data as a reference framework has shown both the potential and the shortcomings of the analysis of such written sources. The data sometimes lack precision regarding geographical provenance and not all taxa received equal attention. In general, it was concluded that the historical data gathered thus far are insufficient to define reference conditions for each water body type. An additional difficulty was the definition of the reference period.[27] This is illustrated by the 'Zeeschelde,' for which it had previously been suggested to set the reference period at AD 1930 because, despite the disappearance or decline of certain species, it was only then that clear signs of ecological stress became visible.[28] At that time, pollution in the estuary was so elevated that fish stocks declined enormously, resulting in an almost complete decline of fisheries. Vrielynck and his colleagues suggested setting the reference period for the 'Zeeschelde' at AD 1900 and believe that the period AD 1900–1930 was a transitional period during which the populations of allis shad (*Alosa alosa*), twaite shad (*Alosa fallax*), smelt (*Osmerus eperlanus*) and flounder (*Platichthys flesus*) were under heavy pressure as shown by their strong fluctuations or tremendous decline.[29]

Even if biological data are used from 1840 onwards, the baselines against which current levels of impact are assessed are not only based on data that already include patterns of severe human disturbance, but also on a time series that is too short (150 years) to be valid within a broad chronological perspective. Historical events, such as climatological changes, periods of war, agricultural revolutions or technological innovations can introduce variation within the dataset that can only be evaluated or smoothed when long time series are available. To document the temporal and spatial patterns in the fish fauna of the Scheldt basin further, it was therefore recommended to investigate additional, and certainly older, historical and archaeological data.[30] This recommendation was especially prompted by the observation that within the Scheldt basin no contemporary unimpacted sites can be found that could serve as a reference. Moreover, these were also not described by the nineteenth- or early twentieth-century biological

---

[27] Ilse Simoens et al., *Monitoringsproject visfauna*.

[28] Dirk Van Damme and Niels De Pauw, *Ontwikkelingsplan voor de visserij op de Schelde beneden Gent. Vakgroep voor toegepaste ecologie en milieubiologie, laboratorium voor biologisch onderzoek van de waterverontreiniging. AMINAL/BNO/WB/VD/94-2, 1996* (Brussel, 1996).

[29] Vrielynck et al., *De visbestanden in Vlaanderen anno 1840–1950*.

[30] Vrielynck et al., *De visbestanden in Vlaanderen anno 1840–1950*.

written sources. Indeed, the Industrial Revolution and subsequent pollution, starting in the middle of the eighteenth century, must have already had a serious impact upon the quality of inland waters. Furthermore, the impact of early modern, medieval, and possibly even Roman industries cannot be neglected. Moreover, it is also known that urbanization, water management works, intensive fishing, and other processes influencing aquatic biotopes and their fish faunas had already manifested themselves early within the historical periods. All of this implies that premodern historical or archaeological information needs to be taken into account when trying to reconstruct minimally disturbed conditions.

### *Unexplored Historical Sources*

Historical documents do indeed corroborate that human disturbance of freshwater ecosystems started well before the nineteenth century.[31] This is not only true for the Scheldt basin, but it also proves to be a general pattern within Europe. The written sources, dating from the late medieval period onwards, document the presence of pollution, water management works and other processes possibly influencing the freshwater fish fauna. Medieval records of pollution due to both organic and inorganic waste in towns are numerous and deal with the disposal in urban environments of domestic refuse and waste water, and with industrial or artisanal waste associated with activities that require water, such as slaughtering, butchering, tanning, brewing, leather-working or wool-dying.[32] The link between this mainly organic pollution and fish mortality is, however, rarely made. The adverse effects of the retting process of flax and hemp on fish populations is sometimes mentioned as, for example, in the 1452 record of a fish kill at Douai, France, for

[31] See, for example, Peter Poulussen, "Ennuis de voisinage et pollution de l'environnement," in *La ville en Flandre: Culture et Société, 1477–1787*, ed. Jan Van der Stock (Brussel, 1991), pp. 71–6; Jean-Pierre Leguay, *La pollution au Moyen Age* (Paris, 1999); Chloé Deligne, "De langetermijngeschiedenis van het afvalbeheer en de watervervuiling in Brussel," in *De geschiedenis van industriële technologie en vervuiling (Jaarboek voor Ecologische Geschiedenis 2003)*, ed. Liesbet van Nieuwenhuyse and Dries Tys (Gent, 2005), pp. 1–16.

[32] See, for example, Françoise Thomas, "Hygiène, approvisionnement en eau et gestion hydrographique à Namur au XV$^{ème}$ siècle," *Annales de la Socité Archéologique de Namur* 68.2 (1994), 235–305; and Leguay, *La pollution au Moyen Age*; and Chloé Deligne, *Bruxelles et sa rivière: Genèse d'un territoire urbain (12$^{e}$–18$^{e}$ siècle)* (Turnhout, 2003).

which the clothier Barru Lourdel was taken to court.[33] It also appears that the link between metal pollution and fish mortality was also rarely made. A notable exception is the sixteenth-century quote of Georgius Agricola in his treatise on mining and metallurgy, stipulating that: "Further, when the ores are washed, the water which has been used poisons the brooks and streams, and either destroys the fish or drives them away."[34] In general, it must be concluded that the effects of pollution on fisheries, well documented from recent studies, must also in medieval times have led to fish mortality and extinction and the development of undesirable and low-quality fish stocks.[35]

As far as freshwater fisheries in Europe are concerned, the most important contribution of historical research regarding human impact is Hoffmann's work about European fisheries since medieval times. Evidence from archives, combined with archaeozoological finds, allowed a detailed reconstruction of the spread of carp breeding (*Cyprinus carpio*) and in more recent papers the author presents historical information documenting the effects of the overexploitation of freshwater fish stocks linked with the growing urban populations and the increased demand for fish for religious reasons (fasting).[36] Human interference thus started well before 'scientific' times. Therefore, for the Scheldt basin, a logical step indeed would be to include historical information dating back before 1840 AD as part of the construction of reference conditions. Certainly, this is possible for the Scheldt basin since a

---

[33] Philippe Plouchard, "La Scarpe et les gens de rivière," in *Fleuves, rivières et canaux dans l'Europe occidentale et médiane*, ed. Pierre Racine (Nancy, 1997), pp. 850–1.

[34] Georgius Agricola, *De re metallica: Translated from the First Latin Edition of 1556 with Biographical Introduction, Annotations and Appendices upon the Development of Mining Methods, Metallurgical Processes, Geology, Mineralogy and Mining Law from the Earliest Times to the 16th Century by Herbert Clark Hoover and Lou Henry Hoover* (London, 1912).

[35] Reinhard Dallinger et al., "Contaminated Food and Uptake of Heavy Metals by Fish: A Review and a Proposal for Further Research," *Oecologia* 73 (1987), 91–8.

[36] Richard C. Hoffmann, "Remains and Verbal Evidence of Carp (*Cyprinus carpio*) in Medieval Europe," in *Fish Exploitation in the Past: Proceedings of the 7th Meeting of the ICAZ Fish Remains Working Group*, ed. Wim Van Neer (Tervuren, 1994), pp. 139–50; idem, "Economic Development and Aquatic Ecosystemes in Medieval Europe," *American Historical Review* 101 (1996), 631–69; idem, "Medieval Fishing," in *Working with Water in Medieval Europe: Technology and Resource Use*, ed. Paolo Squatriti (Leiden, 2000), pp. 331–93; idem, "Frontier Foods for Late Medieval Consumers: Culture, Economy, Ecology," *Environmental History* 7 (2001), 131–67; and idem, "A Brief History of Aquatic Resource Use in Medieval Europe," *Helgoland Marine Research* 59 (2005), 22–30.

wealth of historical documents is available, which is varied in terms of both chronology and social context.[37] In theory, information about freshwater fish can be found in fiscal sources, such as the revenues of powerful families, the toll tariffs for fish traded on the rivers or passing through the city gates, or the tax and price lists for selling fish on the urban markets or in a local fish hall. Freshwater fish are also mentioned in juridical documents, for example, those describing feudal rights or trials as the results of conflicts about water management. Furthermore, fish appear in the set of rules organizing the work of the fishmongers' guild, in contracts delimiting the use and management of fishponds, and in local proscriptions. Exceptional events dealing with catches or occurrences of freshwater fish may have been described in chronicles or diaries. On the consumers' side, fish are present in the bookkeepings of the kitchens of rich households. Exceptional information is provided by iconographic sources such as still-lifes of fish from the sixteenth century and later. In general, the archives of abbeys seem to be the most promising historical sources; these communities took more care in administration than most other parts of society, and they often provide a continuity of evidence (sometimes starting in the early medieval period and extending into recent times).

Unfortunately, the richness in historical sources not only presents a broad spectrum of interpretative possibilities but also a problem of overview and synthesis. At present, no global compilation about the former freshwater fisheries within the Scheldt basin has yet been made for the period before the nineteenth and twentieth centuries.[38] Multiple historical studies have indeed covered older periods but they mainly focused on a limited geographical area (typically a village or a micro-region). An additional problem is that such studies often have a very limited distribution. Students' theses and project reports similarly contain valuable data and insights but they cannot be easily consulted. It would thus be a major enterprise to bring together all the available secondary information (scholarly publications about primary, 'original', historical sources), especially covering the period before the nineteenth century, and evaluate its relevance for the construction of reference conditions.

---

[37] See, for example, Raoul C. Van Caenegem, *Guide to the Sources of Medieval History* (Amsterdam and New York, 1978).

[38] See Vrielynck et al., *De visbestanden in Vlaanderen anno 1840–1950*, and the literature cited there.

Similarly, a renewed study of all primary sources in search of information about the former freshwater fish faunas should be undertaken, but this also will be an enormous task. Typically, the oldest original Flemish historical sources (primarily high and late medieval documents) have been studied over and over (although not always specifically from the viewpoint of environmental history). The much larger corpus of post-medieval sources, however, still remains largely unexplored. It is therefore unknown what types of sources contain unexploited information, what archives still should be consulted, where potentially useful documents are stored, and how this analysis should be organised. But this enterprise is the only way to close the gap between the medieval information and the existing survey of nineteenth- and twentieth-century historical data. From a renewed study of the historical sources a dataset could be extracted, listing the presence of species through time, evaluating the relative abundance or scarcity of species or freshwater fish in general, and describing fishing techniques or water management practices. Taken together, this information will certainly add to the validity of the reference frameworks, be it that a new index will have to be developed, taking into account the specific nature of the information from older historical sources.

### *The Available Archaeological Sources*

Another possible source of information about former freshwater fish faunas consists of the skeletal remains excavated from former human activity sites (settlements, industrial sites, ritual places, and so on), mostly from contexts of consumption refuse. Although limited trade of freshwater fish cannot always be excluded, the freshwater taxa found at a particular archaeological site can usually be considered to represent local catches.[39] Of course, archaeological assemblages do not present a neutral sample taken from a former freshwater fish fauna. The finds' collections are heavily biased by fish catching options and possibilities, market economies, and consumer behavior. From the excavated remains, however, it can still be appreciated whether species were present or not, and possibly what their relative abundance was. Furthermore, demographic parameters recorded from the archaeological material, such as biometric data, evaluations of growth rate, or species age structures, can

[39] Hoffmann, "A Brief History of Aquatic Resource Use in Medieval Europe," p. 23.

serve as indicators of human impact upon fish faunas. Of course, in order to incorporate these data and to take into account their specific nature, a new fish index must again be developed.

The potential of fish remains from archaeological excavations has thus far not frequently been recognized as an additional source for the study of fish ecology from both recent and older periods. In Europe, anecdotal evidence is available for the Loire and fish remains demonstrate the human impact on sturgeon stocks in the Elbe between the sixth and eighth centuries AD.[40] The only instance in which this type of data has been used within a context of sustainable development is a study on the Elbe and Oder reference conditions.[41] The possibilities of archaeological fish bones in differentiating human and climatic impact have recently been demonstrated by a genetic study on sturgeon bones.[42] In Northern Europe, fish bones from Swedish Late Glacial and early Holocene sites have been used to document the post-glacial colonization of the area by freshwater taxa.[43] This type of information is useful for the establishment of the autochthonous part of the ichthyofauna, and prehistoric and early historic fish bones have similarly contributed to the knowledge of the status of present-day species.[44] But detailed studies of the human impact on a single river basin, using a combination of historical, archaeozoological and modern fishery data have thus far not been undertaken.

---

[40] Marie-Christine Marinval, "Les poissons migrateurs du bassin versant de la Loire (France): Bilan préliminaire de leur histoire commune avec les sociétés humaines, de la fin de la dernière glaciation jusqu'à nos jours," in *La migration des animaux: Connaissances zoologiques et exploitations anthropologiques selon les espèces, les lieux et les époques (Colloques d'Histoire des connaissances zoologiques 15)*, ed. Liliane Bodson (Liège, 2004), pp. 61–94; and Norbert Benecke, "Some Remarks on the Sturgeon Fishing in the Southern Baltic Region in Medieval Times," in *Fish and Archaeology: Studies in Osteometry, Taphonomy, Seasonality and Fishing Methods*, ed. Dick C. Brinkhuizen and Anneke T. Clason, British Archaeological Reports International Series 294 (Oxford, 1986), pp. 9–17.

[41] Christian Wolter et al., "The Use of Historical Data to Characterize Fish-Faunistic Reference Conditions for Large Lowland Rivers in Northern Germany," *Archiv für Hydrobiologie Supplement* 155.1–4 (2005), 37–51.

[42] Arne Ludwig et al., "When the American Sea Sturgeon Swam East," *Nature* 419 (2002), pp. 447–8.

[43] Johannes Lepiksaar, *Die spät- und postglaziale Faunengeschichte des Süsswasserfische Schwedens: Übersicht der subfossilen Funde und Versuch einer faunengeschichtlichen Analyse der rezenten Artareale* (Kiel, 2001).

[44] Dick C. Brinkhuizen, "On the Finds of European Catfish (*Siluris glanis* L.) in the Netherlands," in *Archaeozoology I: Proceedings of the IIIrd International Archaeozoological Conference held 23–26th April 1978*, ed. Marian Kubasiewicz (Szczecin, 1979), pp. 256–61.

Regarding the Scheldt basin, fish bones have mainly been used to document archaeological and historical aspects concerning the role of fish as part of the food provisioning, to reconstruct fishing techniques and to establish the status of the consumers, but ecological consequences have received less attention.[45] Indeed, although freshwater fish bones from archaeological sites along the Scheldt have a large potential for biodiversity, zoogeography and fishery biology, these aspects have thus far not yet been exploited in detail. Preliminary data already indicate, however, that the decline of the ichthyofauna started much earlier than the nineteenth century, as studies of recent ecology have traditionally put forward. In this way, the archaeological data certainly corroborate the information from pre-nineteenth-century historical sources. It is clear, for example, from previous work on archaeological fish remains that, through time, a decline in anadromous species occurred. In addition, archaeological fish bone data have clarified the autochthonous status of a number of species. The latter aspect was recognised in the overview of Vrielynck and his colleagues who list the taxa identified thus far from archaeological sites along the Scheldt basin and who use their presence as an argument for the autochthonous status of certain species.[46] A good example of the contribution of archaeoichthyological data to modern fisheries is provided by the archaeological record of the European catfish (*Silurus glanis*), which in some fishery literature is considered to be an allochthonous species for the Scheldt.[47] The species is known from the Scheldt basin since Neolithic times.[48] Historical finds are known from a

---

[45] Wim Van Neer and Anton Ervynck, *Archeologie en vis (Herlevend Verleden 1)* (Zellik, 1993); Wim Van Neer and Anton Ervynck, "New Data on Fish Remains from Belgian Archaeological Sites," in *Fish Exploitation in the Past: Proceedings of the 7th Meeting of the ICAZ Fish Remains Working Group (Annales du Musée Royal de l'Afrique Centrale, Sciences Zoologiques 274)*, ed. Wim Van Neer (Tervuren, 1994), pp. 217–29; and Wim Van Neer and Anton Ervynck, "Remains of Traded Fish in Archaeological Sites: Indicators of Status or Bulk Food?" in *Behaviour Behind Bones: The Zooarchaeology of Ritual, Religion, Status and Identity*, ed. Sharyn Jones O'Day et al. (Oxford, 2004), pp. 203–14.

[46] Vrielynck et al., *De visbestanden in Vlaanderen anno 1840–1950.*

[47] Gerald Louette et al., *Ontwikkeling van een gedocumenteerde gegevensbank over uitheemse vissoorten in Vlaanderen met bijkomend onderzoek naar blauwbandgordel: Eindverslag van project VLINA 00/11 (Studie uitgevoerd voor rekening van de Vlaamse gemeenschap binnen het kader van het Vlaamse Impulsprogramma Natuurontwikkeling in opdracht van de Vlaamse minister bevoegd voor natuurbehoud. D/2002/3241/136)* (Groenendaal, 2001).

[48] Jean Pierre Parent et al., "Prehistorische jagers en veetelers aan de Donk te Oudenaarde," *Vereniging voor Oudheidkundig Bodemonderzoek in Oost-Vlaanderen-info* 24–25 (1987), 1–45.

ninth- and tenth-century context at Tournai, from a 1000 AD assemblage in the *castrum* of Ename, and from the twelfth-century fortification of Seneca at Grimbergen.[49] Also in the Meuse basin, the only other major river basin in Belgium, the fossil record of *Silurus glanis* extends from Neolithic to late medieval times. In that basin the youngest find dates to the fifteenth century AD.[50] Post-medieval finds of the European catfish are also lacking in The Netherlands and in Germany except for a few isolated relic populations.[51] The disappearance of the species is probably a result of over-fishing of an animal with a slow reproduction rate combined with the climatic deterioration during the Little Ice Age. Spawning in shallow, inshore areas would only occur when water temperatures rise to at least 18–20 °C or even 20–22 °C.[52] The specimens that have been captured occasionally in the Scheldt basin in the last two decennia are all believed to represent animals that escaped from experimental fisheries or that were introduced on purpose.[53] Other species that may have been under severe fishing pressure during historical times are houting (*Coregonus oxyrinchus*), salmon (*Salmo salar*), sturgeon (*Acipenser sturio*) and burbot (*Lota lota*). The number of archaeological sites on which these fish are found is limited and the amount of their bones is also always very low. It is unclear if this reflects low natural population densities, a disinterest in these fish (very unlikely in the case of sturgeon) or rather the sensitivity of these species to anthropogenic pressure. The latter factor could be an explanation for the sturgeon: in

---

[49] Tournai: Van Neer, unpublished data from 'site des Cloîtres'; Raymond Brulet et al., "Les sites à «terres noires» à Tournai et l'étude des anciens cloîtres canoniaux: Etudes archéozoologique, palynologique et contextualisation," in *Terres noires: Actes de la table ronde de Louvain-la-Neuve, 09–10 novembre 2001, Département d'archéologie et d'Histoire de l'Art et Centre de Recherches d'Archéologie Nationale (Collection d'Archéologie Joseph Mertens, XIV)*, ed. Raymond Brulet and Laurent Verslype (Louvain-la-Neuve, 2005), pp. 152–72. Ename: Van Neer, unpublished data. Grimbergen: Achilles Gautier and Victor Rubberechts, "Animal Remains of the Senecaberg fortification," *Bulletin Musées royaux d'Art et d'Histoire* 48 (1978), 51–84.

[50] Wim Van Neer and An Lentacker, "Restes fauniques provenant de trois fosses d'aisances du Grognon à Namur (XIIème, XVème–XVIème et XVIIème siècles)," in *Actes de la Quatrième Journée d'Archéologie Namuroise*, ed. Jean Plumier and Marie Hélène Corbiau (Namur, 1996), pp. 89–104.

[51] Brinkhuizen, "On the Finds of European Catfish (*Siluris glanis* L.) in the Netherlands."; and Dirk Heinrich, "Fischreste als archäozoologische Quellengattung – Probleme und Ergebnisse," *Archäologische Informationen* 12.2 (1989), 172–9.

[52] Erna Mohr, *Der Wels* (Wittenberg, 1957); and M. M. Shikhshabekov, "Sexual Cycles of the Catfish, *Silurus glanis*, the Pike, *Esox lucius*, the Perch, *Perca fluviatilis*, and the Pikeperch, *Lucioperca lucioperca*," *Journal of Ichthyology* 18 (1978), 457–68.

[53] Vrielynck et al., *De visbestanden in Vlaanderen anno 1840–1950*.

the Elbe a decline of this species is already noted in early medieval times, a period for which archaeological sites in the Scheldt basin are almost completely lacking.[54] In any case, the available data do not allow us to document at this moment a clear diachronic change in the occurrences.

Another possible indicator of over-fishing, besides a decrease in the numbers of sensitive species, is a decrease in the average size of the freshwater fish. The impression arose from the analysis of a large number of medieval sites in the countryside and in urban contexts that freshwater fish in the latter seem smaller on average. No detailed, osteometric study has thus far been carried out, however, to confirm this impression. Indirect indications for a high pressure on the local freshwater fauna is the breeding of carp and the massive import of marine species.[55] A simultaneous quantification of the proportions of local river fish, carp and marine species may thus be an additional possible measure.

Finally, it must be realized that archaeological information can also provide indications about the processes leading to the disturbance of freshwater fish faunas. On a large scale, archaeology documents the ecological changes in the environment starting with the introduction of agriculture in our part of the world. From the Neolithic period on, deforestation gradually became significant, leading to changes in groundwater levels, soil erosion, and the silting up of rivers. Without doubt, these processes must have had their impact upon the freshwater fish faunas. Archaeozoological material can also bear witness to pollution, as shown by a recent pilot study on goat bones from Sagalassos (Turkey), a Roman to Early Byzantine site (first to seventh century AD) known for its industrial activities.[56] Geochemical analyses of ancient sediments have also occasionally enabled the documentation of the effects of pollution related to ore exploitation, but these studies dealt with terrestrial environments.[57] For the moment, data on pollutant levels in rivers are

---

[54] Benecke, "Some Remarks on the Sturgeon Fishing in the Southern Baltic Region in Medieval Times."

[55] Anton Ervynck and Wim Van Neer, "De overexploitatie van dierlijke grondstoffen uit de vrije natuur: Archeologische indicatoren voor historische fenomenen" *Jaarboek voor Ecologische Geschiedenis* 2004 (2005), 1–18.

[56] Patrick Degryse et al., "Statistical Treatment of Trace Element Data from Modern and Ancient Animal Bone: Evaluation of Roman and Byzantine Environmental Pollution," *Analytical Letters* 37.13 (2004), 1–17.

[57] Pete D. Marshal, "Reconstructing the Environmental Impact of Past Metallurgical Activities," and Varyl R. Thorndycraft et al., "An Environmental Approach to the Archaeology of Tin Mining on Dartmoor," in *The Environmental Archaeology of Industry (Symposia of the Association for Environmental Archaeology 20)*, ed. Peter Murphy and Patricia E. J. Wiltshire (Oxford, 2003), pp. 10–18 and 19–28, respectively.

still lacking. Alternatively, a possible way of documenting the ancient pollution from archaeozoological material may be the proportion of sensitive versus tolerant species. Certain cyprinids, such as roach (*Rutilus rutilus*) or bream (*Abramis brama*) are typical of eutrophic waters, and another possible good indication for poor water quality may be the proportion of eel, a species that is very resistant to both organic and inorganic pollution.[58]

### *Discussion: Integration, Possibilities and Shortcomings*

Working with information from different disciplines, from the humanities to the sciences, is certainly beneficial, because archaeozoological data can cover periods with poor historical information. In addition, a calibration of the various datasets is possible for the later period where there is a temporal overlap between archaeological information and historical data. For the Scheldt basin, a period of approximately 2000 years could be covered, which is much longer than traditional historical methods allow. Of course, the integration of biological, historical and archaeological information must not be undertaken without an assessment of the strengths and weaknesses of the datasets. Promising as it is, the historical record certainly has its limitations. It should be realised that the oldest, useful historical texts, within the context of Flanders, only date back to the High Middle Ages (roughly to the twelfth century AD). Written sources from before that date hardly mention biological or ecological information and certainly do not provide details about fish species. Nevertheless, anthropogenic alterations to the aquatic ecosystems must have started to develop before the earliest period from which meaningful written sources survive.

Fortunately, while scientific biological research can cover only the nineteenth to twenty-first centuries and historical sources do not reach beyond the last millennium, archaeological data can go back to the beginning of the Holocene, and even to times before. Unique to the Scheldt basin is the exceptionally good temporal and spatial

---

58 Christian Wolter et al., "Long-Term Effects of Human Influence on Fish Community Structure and Fisheries in Berlin Waters: An Urban Water System," *Fisheries Management and Ecology* 7 (2000), 97–104; Frederich Tesch, *The Eel: Biology and Management of Anguillid Eels* (Edinburgh, 1977); and Michael J. Hinton and Arnold G. Eversole, "Toxicity and Tolerance Studies with Yellow-Phase Eels: Five Chemicals," *Progressive Fish-Culturist* 42 (1980), 201–3.

coverage of archaeological sites. Since 1990, excavations in Flanders involve the systematic use of large scale, detailed sediment sieving which guarantees the recovery of even the tiniest fish bones. As a result, a massive amount of well-dated fish bones are available for analysis. These archaeological data can complement the historical sources and provide a unique source of information for earlier periods, for which less or no historical sources are available at all. Of course, the archaeological dataset also has its limitations. Although archaeological material from the Scheldt basin is available dating from the Neolithic (ca. 4900–4600 BC) to the beginning of the twentieth century AD, most of the bone assemblages belong to the Roman (first–fourth c. AD), the high and late medieval (ninth–fifteenth centuries AD) and the post-medieval period.[59] This uneven chronological distribution of finds is due to preservation conditions and to the low archaeological visibility of sites from certain historic periods, such as the Early Middle Ages.

In geographical terms, the sites are distributed such that they should allow documenting former fish populations in the Scheldt estuary and in the bream zone of several rivers of the Scheldt basin. Due to the low number of studied archaeological sites in the barbel zone of the Scheldt, information about fish stocks in that part of the basin will need to be obtained mainly through historical analysis. The definition of reference conditions for individual water body types was seriously hampered due to the uneven spread of data over the whole river basin and the vague geographical localization in the ichthyological literature. The most valuable information comes from authors that list precise localities or that deal with the fish fauna of a small region.[60] But most of the fishery reports from the nineteenth and early twentieth century treat river basins as a whole and, when they are more specific, rivers other than the Scheldt, Meuse, Lys or the River Demer are rarely mentioned. Historical and archaeozoological data may allow us to obtain a more detailed picture.

The exploitation of historical sources suffers sometimes from unclear taxonomic nomenclature of species. It appears that, as is still the case

---

[59] Van Neer and Ervynck, *Archeologie en vis*; and Van Neer and Ervynck, "New Data on Fish Remains from Belgian Archaeological Sites."

[60] For example, Max Poll, "Contribution à la connaissance de la faune ichthyologique du Bas-Escaut," and idem, *Poissons marins*; or Constant Bamps and Emile Geraets, *Faune des poissons de la province de Limbourg* (Hasselt, 1897).

today, numerous vernacular names existed in the past to indicate the same species; as a result, sometimes the species that is meant in the texts cannot be precisely identified. Moreover, in certain texts fish names are used that refer to a group of species (for example 'loaches,' 'whitefish'). Such shortcomings also occur occasionally in the ichthyological literature. When de Selys-Longchamps mentions 'loches,' he refers to three species, and it appears that he designates those loaches elsewhere also as 'lottes,' a term that is also used as a vernacular name for the burbot (*Lota lota*).[61] In the case of archaeological material, identification problems are completely different. On the one hand, even isolated fish bones often bear diagnostic features that allow a specific identification.[62] On the other hand, some species groups (like cyprinids or flatfish) are difficult to discriminate on the basis of their skeletal material, but the number of studies dealing with the comparative osteology of individual bones is growing.[63]

Next to identification problems, other factors influence the information per species. Uneven documentation of species in historical records indeed presents a problem. Fish that receive the most attention in the historical documents thus far investigated are rare species regarded as a curiosity (like exotics), fish that were economically important game fish (like salmonids) and high status species (like sturgeon). In archaeological assemblages, there is a bias because the majority of the fish remains represent leftovers of food items. Moreover, the fish spectrum found at an archaeological site reflects the social status of the consumers and the fishing techniques used. Nevertheless, it is striking that numerous sites comprise species of small size that would nowadays be considered as hardly edible, like sticklebacks (Gasterosteidae), loaches (Cobitidae), bullhead (*Cottus gobio*), small-sized cyprinids such as gudgeon (*Gobio gobio*) and minnow (*Phoxinus phoxinus*). This implies that archaeological assemblages are in fact more useful as biological samples than is generally recognised.

---

[61] de Selys-Longchamps, "Classe IV, Poissons d'eau douce."

[62] For example, Roland M. Libois et al., *Eléments pour l'identification des restes crâniens des poissons dulçaquicoles de Belgique et du nord de la France, 1: Anguilliformes, Gastérostéiformes, Cyprinodontiformes et Perciformes (Fiches descriptives animale pour l'archéologie: Série A.: Poissons n°3. Centre de Recherches Archéologiques du CNRS)* (Juan-les-Pins, 1987).

[63] For example, Wim Wouters et al., "The Distinction of Isolated Bones from Plaice (*Pleuronectes platessa*), Flounder (*Platichthys flesus*) and Dab (*Limanda limanda*): A Description of the Diagnostic Characters," *Archaeofauna* 16 (2007), 33–95.

Considering quantification of data, the abundance or rarity of species are only scarcely documented in the written sources. The historical data provide evidence for the occurrence of species, but the proportion of the various species or the ratio of the various age groups within one species are almost always lacking. Notable exceptions are the reports on stocking in ponds written down in the chronicles of large estates. In contrast, archaeological fish bones allow a quantification of the various taxa represented and, in addition, they permit us to reconstruct fish sizes, while age structure and growth patterns can be studied as well.

Of the limitations inherent to the use of archaeological fish bones, the most important one is perhaps differential preservation. Not all species and not all skeletal elements of a particular species have the same chances of surviving the destructive processes that take place between the moment that the fish bones are discarded, and the moment that they are recovered during excavation. This explains why certain taxa have limited chances of being documented archaeozoologically. This is the case for lampreys of which the keratinized 'teeth' are the only elements that can survive in exceptional 'archaeological' circumstances.[64] In Flanders, lamprey has thus far only been documented from Lafelt, a Roman site.[65] Species that have a porous bone structure combined with a high fat content, such as salmonids (including houting), have less survival chances than fish with a well-ossified skeleton. Although these factors hamper the reconstruction of the exact frequency of taxa, it is important to realise that diachronic trends can still be detected as the structural integrity of the bones of the various species remains constant, regardless their archaeological context. It is common practice in archaeozoological analysis to establish proportions between taxa and to explain the observed trends in these ratios within a palaeo-ecological or palaeo-economical framework.[66]

---

[64] Andrew K. G. Jones, "Provisional remarks on fish remains from archaeological deposits at York," in *The Exploitation of Wetlands*, ed. Peter Murphy and Charles French, British Archaeological Reports, British Series 186 (Oxford, 1988), pp. 113–27; and Ruby Ceron Carrasco, "Fish and Crustaceans," in *Dundrennan Abbey: Archaeological Investigation Within the South Range of a Cistercian House in Kirkcudbrightshire (Dumfries & Galloway), Scotland*, ed. Gordon Ewart (Scottish Archaeological Internet Reports 1, 2001): http://www.sair.org.uk/sair1/index.html

[65] Wim Van Neer and Anton Ervynck, unpublished data.

[66] Achilles Gautier, "How Do I Count You, Let Me Count the Ways? Problems of Archaeozoological Quantification," in *Animals and Archaeology 4: Husbandry in Europe*, ed. Caroline Grigson and Juliet Clutton-Brock, British Archaeological Reports International Series 227 (Oxford, 1984), pp. 237–51.

This approach has been successful in the establishment of diachronic trends in marine exploitation.[67]

Assuming that historical research that will eventually compile a coherent corpus of data on freshwater fish and archaeological analyses can continue, together with the ichthyological literature these datasets will provide a useful basis for the establishment of reference conditions. At the level of individual species, occurrences will have to be evaluated through time, thus providing evidence about their status (autochthonous or introduced) as well as about effects of human and/or climatic impact on their abundance (and in certain cases, extinction). From the archaeological data, for each taxon, the trends seen in the size distributions will have to be evaluated, such as possible decrease in size through time or changes in the (normal) distribution of sizes. At the (archaeological) population level, the fragment counts of the various taxa identified can be used to calculate diachronic trends in the ichthyofauna. This will ultimately lead to a dataset that is complementary to the already existing database on (sub)recent fish populations at INBO. Using these data, a list of relevant candidate metrics should be retained that can subsequently be tested for their sensitivity to human impact. From these statistical tests a list of selected metrics should then be compiled. Hence, the interpretation of the datasets should allow us to follow: (1) individual species through time (their presence, their relative abundance), documenting the fate of anadromous and other sensitive species and reconstructing the introduction of species; (2) species spectrum changes; and (3) trends in size distributions through time. These data should then be integrated with present-day fishery data, and, through modeling, contribute to the development of an IBI specifically for archaeological and historical material. Ultimately, it must be possible to use these data to adapt the previously established reference conditions.[68]

### *Conclusion and a Note of Warning*

In general, all of the methodological problems listed above will not outweigh the advantages of building reference frameworks on the basis of

---

67 Anton Ervynck et al., "How the North was Won (and Lost Again): Historical and Archaeological Data on the Exploitation of the North Atlantic by the Flemish Fishery," in *Atlantic Connections and Adaptations: Economies, Environments and Subsistence in Lands Bordering the North Atlantic*, ed. Rupert A. Housley and Geraint Coles (Oxford, 2004), pp. 230–9.

68 Ilse Simoens et al., *Monitoringsproject visfauna*.

an integrated dataset of biological, historical and archaeological information. Surprisingly, however, such projects have not yet been put into practice very often. Such a synthesis still needs to be made for the Scheldt basin and thus far only a few pilot studies have been carried out in other parts of the world. Two examples of an integrated approach are known from Oregon and California in the United States.[69] The use of nitrogen isotopes and diatoms from lake sediment cores, documenting the sockeye salmon (*Oncorhynchus nerka*) abundances on their spawning grounds over the last 2200 years, showed biomass variability and its relationship to climatic conditions in the northeastern Pacific Ocean.[70] Another approach using historical sources is the analysis of detailed mid-nineteenth-century fishing logs with daily catch records and fleet activity in New England that allowed for the estimation of former cod biomass.[71] Although relatively less detailed information is to be expected, ongoing historical research on documents dating back as far as AD 1500 aims at reconstructing former biomass in the Baltic Sea and Skagerrak, taking into account factors such as climatic change, eutrophication and predation.[72] The latter initiative also intends to include information provided by archaeological fish remains. The combined use of historical and archaeo-ichthyological data is indeed believed to be a promising approach to monitor long-term impact of human activities on aquatic ecosystems.[73] For the North Sea, a pilot project has recently been initiated that aims at documenting and understanding the intensification of fisheries by a combination of archaeozoological and chemical analyses of samples dating between 600 and 1600 AD.[74] An analysis of the otoliths

---

[69] Virginia L. Butler and Michael G. Delacorte, "Doing Zooarchaeology As If It Mattered: Use of Faunal Data to Address Current Issues in Fish Conservation Biology in Owens Valley, California," in *Zooarchaeology and Conservation Biology*, ed. R. Lee Lyman and Kenneth P. Cannon (Salt Lake City, 2004), pp. 25–44.

[70] Bruce P. Finney et al., "Fisheries Productivity in the Northeastern Pacific Ocean over the Past 2,200 Years," *Nature* 416 (2002), 729–33.

[71] Andrew A. Rosenberg et al., "The History of Ocean Resources: Modeling Cod Biomass Using Historical Records," *Frontiers in Ecology and the Environment* 3 (2005), 84–90.

[72] Brian R. MacKenzie et al., "Ecological Hypotheses for a Historical Reconstruction of Upper Trophic Level Biomass in the Baltic Sea and Skagerrak," *Canadian Journal of Fisheries and Aquatic Sciences* 59 (2002), 173–90.

[73] Daniel Pauly, "Much Rowing for Fish," *Nature* 432 (2004), 813–4.

[74] James H. Barrett et al., "The Origins of Intensive Marine Fishing in Medieval Europe: The English Evidence," *Proceedings of the Royal Society of London Series B – Biological Sciences* 271 (2004), 2417–2421; and James H. Barrett et al., "Detecting the Medieval Cod Trade: A New Method and First Results," *Journal of Archaeological Science* 35 (2008), 850–61.

of cod (*Gadus morhua*), haddock (*Melanogrammus aeglefinus*) and plaice (*Pleuronectes platessa*) from archaeological sites in Belgium, dating between the fifteenth and twentieth centuries AD and the comparison of their size and growth increments to modern otoliths showed signs of overfishing prior to the onset of industrial fishing in the mid-nineteenth century.[75] This trend could be established through changes in fish size and growth rate. All of the aforementioned analyses and initiatives deal with marine fisheries, whereas the research lines proposed here intend to document long-term changes in the freshwater fish fauna through a combined approach involving historical and archaeozoological data as part of ecological modeling.

From this line of reasoning, it is clear that through the incorporation of historical and archaeological data, reference conditions can be reconstructed that are closer to the original 'natural' condition, prior to significant human disturbance, than those solely based upon the nineteenth- and twentieth-century ichthyological publications. The ultimate 'natural' condition will always stay out of sight, however. It must not be forgotten that all data gathered by definition derive from an anthropogenic context (historical records, archaeological sites) and thus show some human influence. Pure palaeontological data, such as information from fish deposited (in post-glacial times) in a natural way, without direct human interference, are nonexistent in Europe besides a few exceptions, for example, in Sweden.[76] In the Scheldt basin and also elsewhere in Europe virtually all fish bone assemblages are anthropogenic in nature. The analysis of natural death assemblages would also give a biased picture, however, as shown by many taphonomic studies carried out on modern fish.[77] In any case, a return to a purely 'natural' situation will not be possible, given the fact that some human influence

---

75 Wim Van Neer et al., "Fish Otoliths and Their Relevance to Archaeology: An Analysis of Medieval, Post-Medieval and Recent Material of Plaice, Cod and Haddock from the North Sea," *Environmental Archaeology* 7 (2002), 65–81; Loes J. Bolle et al., "Growth Changes in Plaice, Cod, Haddock and Saithe in the North Sea: A Comparison of (Post-)Medieval and Present-Day Growth Rates based on Otolith Measurements," *Journal of Sea Research* 51 (2004), 313–28; and Virginia L. Butler, "Where have All the Native Fish Gone? The Fate of the Fish that Lewis and Clark Encountered on the Lower Columbia River," *Oregon Historical Quarterly* 105 (2004), 438–63.

76 Lepiksaar, *Die spät- und postglaziale Faunengeschichte des Süsswasserfische Schwedens*.

77 Ruth L. Elder and Gerald R. Smith, "Fish Taphonomy and Environmental Inference in Paleolimnology," *Palaeogeography, Palaeoclimatology, Palaeoecology* 62 (1988), 577–92.

will always remain. Regarding nature conservation, the question remains how much human impact we want to tolerate, rather than fanatically trying to rebuild a situation impossible to reach or to maintain.

As an approach to establishing riverine reference conditions, the WFD put forward several alternative options for reference frameworks: (1) a state prior to any human disturbance; (2) a state prior to the generalised introduction of intensive agriculture; (3) a state prior to intensive agriculture, extensive air-born industrial pollution, or significant abstraction (removal of water typically for industrial or agricultural uses) and significant morphological change (and any other significant pressures); or (4) the currently best available state of any water body type in any ecoregion.[78] Taking into account that nature protection and conservation must be more ambitious than only trying to keep what one has, and given the fact that at present almost all waters within the Scheldt basin are heavily disturbed by humans, option 4 is hardly a valuable one. At the same time, options 1 and 2 seem impossible to reach given the extreme human population density in Northwestern Europe. This leaves us with option 3, a situation that must be studied through history and archaeology, following the proposals made earlier in this study. The intensification of agriculture or the rise of pollution levels have all been very gradual processes, however, making it difficult to define when the situation described by option 3 was lost. This implies that it will be better not to take a particular period *a priori* as a temporal benchmark. In fact, the aim must be to study the evolution of the fish fauna going back in time as far as possible and to evaluate within a broad chronological perspective how human interference developed. Then, taking into account all economic options, it can be decided what former conditions would be the most suitable as a reference.

In conclusion, the potential of an integrated approach is enormous. This does not necessarily imply, however, that such an integrated dataset will easily be achieved. A lot of work still needs to be done. The nineteenth- and twentieth-century ichthyological publications deserve more attention although most of their potential has been explored. The archaeological dataset on freshwater fish from the Scheldt basin is slowly growing. The research is limited by time and budgetary

---

[78] Roger Owen et al., *Definition and Establishment of Reference Conditions.*

constraints but progress is being made. In contrast, the construction of the historical dataset, despite the wealth and variety of the possible information sources, awaits a new impetus. Significant progress will thus be made when historians start to take more interest in ecological themes such as the evolution of the freshwater fauna. We hope that this contribution can serve as an invitation.

## *Bibliography*


Agricola, Georgius. *De re metallica: Translated from the First Latin Edition of 1556 with Biographical Introduction, Annotations and Appendices upon the Development of Mining Methods, Metallurgical Processes, Geology, Mineralogy and Mining Law from the Earliest Times to the 16th Century by Herbert Clark Hoover and Lou Henry Hoover* (London, 1912).

Bamps, Constant and Emile Geraets, *Faune des poissons de la province de Limbourg* (Hasselt, 1897).

Barrett, James H. et al., "The Origins of Intensive Marine Fishing in Medieval Europe: The English Evidence," *Proceedings of the Royal Society of London Series B – Biological Sciences* 271 (2004), 2417–21.

——. "Detecting the Medieval Cod Trade: A New Method and First Results," *Journal of Archaeological Science 35* (2008), 850–861.

Belpaire, Claude et al., "An Index of Biotic Integrity Characterizing Fish Populations and the Ecological Quality of Flandrian Water Bodies," *Hydrobiologia* 434 (2000), 17–33.

Benecke, Norbert. "Some Remarks on the Sturgeon Fishing in the Southern Baltic Region in Medieval Times," in *Fish and Archaeology: Studies in Osteometry, Taphonomy, Seasonality and Fishing Methods*, ed. Dick C. Brinkhuizen and Anneke T. Clason, British Archaeological Reports International Series 294 (Oxford, 1986), pp. 9–17.

Bolle, Loes J. et al., "Growth Changes in Plaice, Cod, Haddock and Saithe in the North Sea: A Comparison of (Post-)Medieval and Present-Day Growth Rates based on Otolith Measurements," *Journal of Sea Research* 51 (2004), 313–28.

Breine, Jan et al., "A Fish-Based Index of Biotic Integrity for Upstream Brooks in Flanders (Belgium)," *Hydrobiologia* 522 (2004), 133–48.

——. "A Fish-Based Assessment Tool for the Ecological Quality of the Brackish Schelde Estuary in Flanders (Belgium)," *Hydrobiologia* 575 (2007), 141–59.

Brinkhuizen, Dick C. "On the Finds of European Catfish (*Siluris glanis* L.) in the Netherlands," in *Archaeozoology I: Proceedings of the IIIrd International Archaeozoological Conference held 23-26th April 1978*, ed. Marian Kubasiewicz (Szczecin, 1979), pp. 256–61.

Brulet, Raymond et al., "Les sites à «terres noires» à Tournai et l'étude des anciens cloîtres canoniaux: Etudes archéozoologique, palynologique et contextualisation," in *Terres noires: Actes de la table ronde de Louvain-la-Neuve, 09-10 novembre 2001, Département d'archéologie et d'Histoire de l'Art et Centre de Recherches d'Archéologie Nationale (Collection d'Archéologie Joseph Mertens, XIV)*, ed. Raymond Brulet and Laurent Verslype (Louvain-la-Neuve, 2005), pp. 152–72.

Butler, Virginia L. "Where have All the Native Fish Gone? The Fate of the Fish that Lewis and Clark Encountered on the Lower Columbia River," *Oregon Historical Quarterly* 105 (2004), 438–63.

Butler, Virginia L. and Michael G. Delacorte, "Doing Zooarchaeology As If It Mattered: Use of Faunal Data to Address Current Issues in Fish Conservation Biology in

Owens Valley, California," in *Zooarchaeology and Conservation Biology*, ed. R. Lee Lyman and Kenneth P. Cannon (Salt Lake City, 2004), pp. 25–44.

Ceron Carrasco, Ruby. "Fish and Crustaceans," in *Dundrennan Abbey: Archaeological Investigation Within the South Range of a Cistercian House in Kirkcudbrightshire (Dumfries & Galloway), Scotland*, ed. Gordon Ewart (Scottish Archaeological Internet Reports 1, 2001): http://www.sair.org.uk/sair1/index.html

Dallinger, Reinhard et al., "Contaminated Food and Uptake of Heavy Metals by Fish: A Review and a Proposal for Further Research," *Oecologia* 73 (1987), 91–8.

Deligne, Chloé. *Bruxelles et sa rivière: Genèse d'un territoire urbain (12e–18e siècle)* (Turnhout, 2003).

——. "De langetermijngeschiedenis van het afvalbeheer en de watervervuiling in Brussel," in *De geschiedenis van industriële technologie en vervuiling (Jaarboek voor Ecologische Geschiedenis 2003)*, ed. Liesbet van Nieuwenhuyse and Dries Tys (Gent, 2005), pp. 1–16.

Degryse, Patrick et al., "Statistical Treatment of Trace Element Data from Modern and Ancient Animal Bone: Evaluation of Roman and Byzantine Environmental Pollution," *Analytical Letters* 37.13 (2004), 1–17.

Didier, Jacques. *Indice Biotique d'Intégrité Piscicole (IBIP) pour évaluer la qualité écologique des écosystèmes lotiques* (Namur, 1997).

Dillen, Alain et al., *Onderzoek naar de biologie van de kwabaal (Lota lota L.), ter voorbereiding van het herstel van de soort in het Vlaamse Gewest (Rapport van het Instituut voor Natuurbehoud 2005.04)* (Brussel, 2005).

Economou, Alcibiades N. *Development, Evaluation and Implementation of a Standardised Fish-Based Assessment Method for the Ecological Status of European Rivers: A Contribution to the Water Framework Directive (FAME), Defining Reference Conditions (D3), Final Report* (2002); see http://fame.boku.ac.at/downloads/D3_reference_conditions.pdf

Elder, Ruth L. and Gerald R. Smith, "Fish Taphonomy and Environmental Inference in Paleolimnology," *Palaeogeography, Palaeoclimatology, Palaeoecology* 62 (1988), 577–92.

Environmental Protection Agency (EPA), *Metrics and the Index of Biotic Integrity* (2006); see http://www.epa.gov/bioindicators/html/ibi-metrics.html

Ervynck, Anton et al., "How the North was Won (and Lost Again): Historical and Archaeological Data on the Exploitation of the North Atlantic by the Flemish Fishery," in *Atlantic Connections and Adaptations: Economies, Environments and Subsistence in Lands Bordering the North Atlantic*, ed. Rupert A. Housley and Geraint Coles (Oxford, 2004), pp. 230–9.

Ervynck, Anton and Wim Van Neer, "De overexploitatie van dierlijke grondstoffen uit de vrije natuur: Archeologische indicatoren voor historische fenomenen" *Jaarboek voor Ecologische Geschiedenis* 2004 (2005), 1–18.

Finney, Bruce P. et al., "Fisheries Productivity in the Northeastern Pacific Ocean over the Past 2,200 Years," *Nature* 416 (2002), 729–733.

Gautier, Achilles. "How Do I Count You, Let Me Count the Ways? Problems of Archaeozoological Quantification," in *Animals and Archaeology 4: Husbandry in Europe*, ed. Caroline Grigson and Juliet Clutton-Brock, British Archaeological Reports International Series 227 (Oxford, 1984), pp. 237–51.

Gautier, Achilles and Victor Rubberechts, "Animal Remains of the Senecaberg fortification," *Bulletin Musées royaux d'Art et d'Histoire* 48 (1978), 51–84.

Goffaux, Delphine et al., *A Biotic Index of Fish Integrity (IBIP) to Evaluate the Ecological Quality of Lotic Ecosystems: Application to the Meuse River Basin, Final Report (User Guide Life 97 ENV/B/000419)* (Groenendaal, 2001).

Heinrich, Dirk "Fischreste als archäozoologische Quellengattung – Probleme und Ergebnisse," *Archäologische Informationen* 12.2 (1989), 172–9.

Hinton, Michael J. and Arnold G. Eversole, "Toxicity and Tolerance Studies with Yellow-Phase Eels: Five Chemicals," *Progressive Fish-Culturist* 42 (1980), 201–3.

Hoffmann, Richard C. "Remains and Verbal Evidence of Carp (*Cyprinus carpio*) in Medieval Europe," in *Fish Exploitation in the Past: Proceedings of the 7th Meeting of the ICAZ Fish Remains Working Group*, ed. Wim Van Neer (Tervuren, 1994), pp. 139–50.

——. "Economic Development and Aquatic Ecosystemes in Medieval Europe," *American Historical Review* 101 (1996), 631–69.

——. "Medieval Fishing," in *Working with Water in Medieval Europe: Technology and Resource Use*, ed. Paolo Squatriti (Leiden, 2000), pp. 331–93.

——. "Frontier Foods for Late Medieval Consumers: Culture, Economy, Ecology," *Environmental History* 7 (2001), 131–67.

——. "A Brief History of Aquatic Resource Use in Medieval Europe," *Helgoland Marine Research* 59 (2005), 22–30.

Huet, Marcel "Profiles and Biology of Western European Streams as Related to Fish Management," *Transactions of the American Fisheries Society* 88.3 (1959), 155–63.

Jones, Andrew K. G. "Provisional remarks on fish remains from archaeological deposits at York," in *The Exploitation of Wetlands*, ed. Peter Murphy and Charles French, British Archaeological Reports, British Series 186 (Oxford, 1988), pp. 113–27.

Karr, James R. "Assessment of Biotic Integrity Using Fish Communities," *Fisheries (Bethesda)* 6 (1981), 21–7.

Kestemont, Patrick et al., "Selecting Ichthyological Metrics to Assess River Quality Basin Ecological Quality," *Archiv für Hydrobiologie Supplement* 121.3–4 (1999), 321–48.

Lameere, Auguste. *Manuel de la faune de Belgique 1: Animaux non insectes* (Bruxelles, 1895).

Leguay, Jean-Pierre. *La pollution au Moyen Age* (Paris, 1999).

Lepiksaar, Johannes. *Die spät- und postglaziale Faunengeschichte des Süsswasserfische Schwedens: Übersicht der subfossilen Funde und Versuch einer faunengeschichtlichen Analyse der rezenten Artareale* (Kiel, 2001).

Libois, Roland M. et al., *Eléments pour l'identification des restes crâniens des poissons dulçaquicoles de Belgique et du nord de la France, 1: Anguilliformes, Gastérostéiformes, Cyprinodontiformes et Perciformes (Fiches descriptives animale pour l'archéologie: Série A.: Poissons n°3. Centre de Recherches Archéologiques du CNRS)* (Juan-les-Pins, 1987).

Louette, Gerald et al., *Ontwikkeling van een gedocumenteerde gegevensbank over uitheemse vissoorten in Vlaanderen met bijkomend onderzoek naar blauwbandgordel: Eindverslag van project VLINA 00/11 (Studie uitgevoerd voor rekening van de Vlaamse gemeenschap binnen het kader van het Vlaamse Impulsprogramma Natuurontwikkeling in opdracht van de Vlaamse minister bevoegd voor natuurbehoud. D/2002/3241/136)* (Groenendaal, 2001).

Ludwig, Arne et al., "When the American Sea Sturgeon Swam East," *Nature* 419 (2002), pp. 447–8.

MacKenzie, Brian R. et al., "Ecological Hypotheses for a Historical Reconstruction of Upper Trophic Level Biomass in the Baltic Sea and Skagerrak," *Canadian Journal of Fisheries and Aquatic Sciences* 59 (2002), 173–90.

Maes, Joachim et al., "The Composition of the Fish and Crustacean Community of the Zeeschelde Estuary (Belgium)," *Belgian Journal of Zoology* 127 (1997), 47–55.

Maes, L. *Notes sur la pêche fluviale et maritime en Belgique* (Bruxelles, 1898).

Marinval, Marie-Christine. "Les poissons migrateurs du bassin versant de la Loire (France): Bilan préliminaire de leur histoire commune avec les sociétés humaines, de la fin de la dernière glaciation jusqu'à nos jours," in *La migration des animaux: Connaissances zoologiques et exploitations anthropologiques selon les espèces, les lieux et les époques (Colloques d'Histoire des connaissances zoologiques 15)*, ed. Liliane Bodson (Liège, 2004), pp. 61–94.

Marshal, Pete D. "Reconstructing the Environmental Impact of Past Metallurgical Activities," in *The Environmental Archaeology of Industry (Symposia of the Association for Environmental Archaeology 20)*, ed. Peter Murphy and Patricia E. J. Wiltshire (Oxford, 2003), pp. 10–8.

Mohr, Erna. *Der Wels* (Wittenberg, 1957).

Owen, Roger et al., *Definition and Establishment of Reference Conditions: Water Framework Directive, REFCOND, Ispra* (Aberdeen, 2001).

Parent, Jean Pierre et al., "Prehistorische jagers en veetelers aan de Donk te Oudenaarde," *Vereniging voor Oudheidkundig Bodemonderzoek in Oost-Vlaanderen-info* 24–25 (1987), 1–45.

Pauly, Daniel. "Much Rowing for Fish," *Nature* 432 (2004), 813–4.

Plouchard, Philippe. "La Scarpe et les gens de rivière," in *Fleuves, rivières et canaux dans l'Europe occidentale et médiane*, ed. Pierre Racine (Nancy, 1997), pp. 850–1.

Poll, Max. "Contribution à la connaissance de la faune ichthyologique du Bas-Escaut," *Bulletin du Musée Royal d'Histoire Naturelle de Belgique* 21 (1945), 1–32.

——. *Poissons marins* (Brussels, 1947).

Poulussen, Peter. "Ennuis de voisinage et pollution de l'environnement," in *La ville en Flandre: Culture et Société, 1477–1787*, ed. Jan Van der Stock (Brussel, 1991), pp. 71–6.

Raveret-Wattel, Casimir. *Atlas de poche des poissons d'eau douce de la France, de la Suisse Romande et de la Belgique: avec leur description, moeurs et organisation*, Bibliothéque de poche du naturaliste 11 (Paris, 1900).

Rosenberg, Andrew A. et al., "The History of Ocean Resources: Modeling Cod Biomass Using Historical Records," *Frontiers in Ecology and the Environment* 3 (2005), 84–90.

Rousseau, Ernest et al., *La pêche fluviale en Belgique: Manuel à l'usage des pêcheurs et des gardes* (Bruxelles, 1915).

de Selys-Longchamps, Edmond. "Classe IV, Poissons d'eau douce," in *Faune belge 1: Indication méthodique des mammifères, oiseaux, reptiles et poissons observés jusqu'ici en Belgique* (Liège, 1842).

——. *La pêche fluviale en Belgique*, Extrait de l'Académie royale de Belgique, 2ième série, tome 22, n. 12 (Bruxelles, 1867).

——. *Révisions des poissons d'eau douce de la faune Belge*, Extrait de l'Académie royale de Belgique, 3me série, tome 14, n. 12 (Bruxelles: Académie royale de Belgique, 1887).

Shikhshabekov, M. M. "Sexual Cycles of the Catfish, *Silurus glanis*, the Pike, *Esox lucius*, the Perch, *Perca fluviatilis*, and the Pikeperch, *Lucioperca lucioperca*," *Journal of Ichthyology* 18 (1978), 457–68.

Simoens, Ilse et al., *Monitoringsproject visfauna: Afleiden en beschrijven van systeemeigen referentieomstandigheden en/of maximaal ecologisch potentieel voor visgemeenschappen in elk Vlaams oppervlaktewaterlichaamtype, vanuit de – overeenkomstig de Kaderrichtlijn Water – ontwikkelde beoordelingssystemen op basis van vismonitoring (Onderzoeksopdracht nr.: VMM.AMO.SCALDIT.VISII)* (Groenendaal, 2005).

Soetaert, Karline, and Peter M. J. Herman, "Estimating Residence Times in the Westerschelde (The Netherlands) Using a Box Model with Fixed Dispersion Coefficients," in *Major Biological Processes in European Tidal Estuaries*, ed. Carl H. R. Heip and Peter M. J. Herman (London, 1995), pp. 215–24.

Tesch, Frederich. *The Eel: Biology and Management of Anguillid Eels* (Edinburgh, 1977).

Thomas, Françoise. "Hygiène, approvisionnement en eau et gestion hydrographique à Namur au XVème siècle," *Annales de la Socité Archéologique de Namur* 68.2 (1994), 235–305.

Thorndycraft, Varyl R. et al., "An Environmental Approach to the Archaeology of Tin Mining on Dartmoor," in *The Environmental Archaeology of Industry (Symposia of the Association for Environmental Archaeology 20)*, ed. Peter Murphy and Patricia E. J. Wiltshire (Oxford, 2003), pp. 19–28.

Van Aelbroeck, Florent, and Emile Rentiers, *Poissons des eaux douces et saumâtres dans leur habitat* (Bruxelles, 1913).
Van Braeckel, Alexander et al., *Historische analyse van de Zeeschelde en haar getijgebonden zijrivieren, 19e eeuw tot heden: Report INBO.R.2006.29* (Brussel, 2006).
Van Caenegem, Raoul C. *Guide to the Sources of Medieval History* (Amsterdam and New York, 1978).
Van Damme, Dirk, and Niels De Pauw, *Ontwikkelingsplan voor de visserij op de Schelde beneden Gent. Vakgroep voor toegepaste ecologie en milieubiologie, laboratorium voor biologisch onderzoek van de waterverontreiniging. AMINAL/BNO/WB/VD/94-2, 1996* (Brussel, 1996).
Van Damme, Paul A. et al., "Fish Species in the Lower Zeeschelde (Belgium): A Comparison with Historical Checklists," *Belgian Journal of Zoology* 124 (1994), 93–103.
Van Neer, Wim et al., "Fish Otoliths and Their Relevance to Archaeology: An Analysis of Medieval, Post-Medieval and Recent Material of Plaice, Cod and Haddock from the North Sea," *Environmental Archaeology* 7 (2002), 65–81.
Van Neer, Wim and Anton Ervynck. *Archeologie en vis (Herlevend Verleden 1)* (Zellik, 1993).
——. "New Data on Fish Remains from Belgian Archaeological Sites," in *Fish Exploitation in the Past: Proceedings of the 7th Meeting of the ICAZ Fish Remains Working Group (Annales du Musée Royal de l'Afrique Centrale, Sciences Zoologiques 274)*, ed. Wim Van Neer (Tervuren, 1994), pp. 217–29.
——. "Remains of Traded Fish in Archaeological Sites: Indicators of Status or Bulk Food?" in *Behaviour Behind Bones: The Zooarchaeology of Ritual, Religion, Status and Identity*, ed. Sharyn Jones O'Day et al. (Oxford, 2004), pp. 203–14.
Van Neer, Wim and An Lentacker, "Restes fauniques provenant de trois fosses d'aisances du Grognon à Namur (XIIème, XVème-XVIème et XVIIème siècles)," in *Actes de la Quatrième Journée d'Archéologie Namuroise*, ed. Jean Plumier and Marie Hélène Corbiau (Namur, 1996), pp. 89–104.
Van Strydonk, Mark, and Guy De Mulder, *De Schelde. Verhaal van een rivier* (Leuven, 2000).
Vandelannoote, Alain et al., *Atlas van de Vlaamse beek- en riviervissen* (Wijnegem, 1998).
Vrielynck, Sven et al., *De visbestanden in Vlaanderen anno 1840–1950: Een historische schets van de referentietoestand van onze waterlopen aan de hand van de visstand, ingevoerd in een databank en vergeleken met de actuele toestand. (Rapport Instituut voor Bosbouw en Wildbeheer IBW.Wb.V.R.2002.89)* (Hoeilaart, 2003).
Wallin, Mats et al., *Final Guidance on Establishing Reference Conditions and Ecological Status Class Boundaries for Inland Surface Waters: EU Common Implementation Strategy (CIS) for the Water Framework Directive* (Luxemburg, 2003).
Wolter, Christian et al., "Long-Term Effects of Human Influence on Fish Community Structure and Fisheries in Berlin Waters: An Urban Water System," *Fisheries Management and Ecology* 7 (2000), 97–104.
——. "The Use of Historical Data to Characterize Fish-Faunistic Reference Conditions for Large Lowland Rivers in Northern Germany," *Archiv für Hydrobiologie Supplement* 155.1–4 (2005), 37–51.
Wouters, Wim et al., "The Distinction of Isolated Bones from Plaice (*Pleuronectes platessa*), Flounder (*Platichthys flesus*) and Dab (*Limanda limanda*): A Description of the Diagnostic Characters," *Archaeofauna* 16 (2007), 33–95.

# APPENDIX: RICHARD C. HOFFMANN: WORKS PUBLISHED, 1971–2008

Compiled by Scott G. Bruce with the help of Ellen Hoffmann and Jean Levy

1971

(with H. B. Johnson, Jr.) "Un village portugais en mutation: Povóa d'El Rey a la fin du XIVe siecle," *Annales: économies, sociétés, civilizations* 16 (1971), 917–40.

1973

"Warfare, Weather, and a Rural Economy: The Duchy of Wrocław in the Mid-Fifteenth Century," *Viator* 4 (1973), 273–305.

1974

"Nazwy i miejscowości: Trzy studia z historii średniowiecznej okręgu wrocławskiego (Place Names and Places: Three Studies in the Medieval Wrocław District)," *Ślęski Kwartalnik Historyczny. Sobótka* 29 (1974), 1–25.

1975

"Medieval Origins of the Common Fields," in *European Peasants and their Markets*, ed. William N. Parker and Eric Jones (Princeton, 1975), pp. 23–71.

"Towards a 'City-State' in East-Central Europe: Control of Local Government in the Late Medieval Duchy of Wrocław," *Societas: A Review of Social History* 5 (1975), 173–99.

1977

"Wrocław Citizens as Rural Landholders," in *The Medieval City*, ed. H. Miskimin, D. Herlihy, and A. L. Udovitch (New Haven, 1977), pp. 293–312.

1982

"A New Treatise on the Treatyse," *The American Fly Fisher* 9:3 (Summer 1982), pp. 2–6.

1983

"Outsiders by Birth and Blood: Racist Ideologies and Realities around the Periphery of Medieval European Culture," *Studies in Medieval and*

*Renaissance History* 6 (1983), 3–34; reprinted in *The Medieval Frontiers of Latin Christendom: Expansion, Contraction, Continuity*, ed. James Muldoon and Felipe Fernández-Armesto (Burlington, VT, 2008), pp. 149–80.

1984

(trans. with T. V. Cohen) "El Tratadico de la Pesca: The Little Treatise on Fishing of Fernando Basurto … 1539," *The American Fly Fisher* 11:3 (Summer, 1984), pp. 8–13.

"The Evidence for Early European Angling, I: Basurto's *Dialogo* of 1539," *The American Fly Fisher* 11:4 (Fall, 1984), pp. 2–9.

"Undertaking a Rehabilitation Project," in Ontario Ministry of Natural Resources, *Community Fisheries Involvement Program Field Manual, Part 1: Trout Stream Rehabilitation* (Toronto, n.d. [1984]), pp. 35–42.

1985

"Fish Ponds," in *Dictionary of the Middle Ages*, ed. Joseph R. Strayer, 12 vols. (New York, 1985), vol. 5, pp. 73–4.

"Fishing for Sport in Medieval Europe: New Evidence," *Speculum* 60 (1985), 877–902.

1987

"The Protohistory of Pike in Western Culture," in *An Annotated Bibliography of the Pike Esox lucius*, ed. E. J. Crossman and J. Casselman, Life Sciences Miscellaneous Publications (Toronto, 1987), pp. vii–xvi; reprinted in *The Medieval World of Nature: A Book of Essays*, ed. J. E. Salisbury (New York, 1993), pp. 61–76.

1988

"Tenure of Land, Western European," in *Dictionary of the Middle Ages*, ed. Joseph R. Strayer, 12 vols. (New York, 1988), vol. 11, pp. 671–86.

1989

*Land, Liberties, and Lordship in a Late Medieval Countryside: Agrarian Structures and Change in the Duchy of Wrocław* (Philadelphia, 1989) [xx + 578 pp.].

"Tools, Agricultural: European," in *Dictionary of the Middle Ages*, ed. Joseph R. Strayer, 12 vols. (New York, 1989), vol. 12, pp. 72–83.

"Villages, Community," in *Dictionary of the Middle Ages*, ed. Joseph R. Strayer, 12 vols. (New York, 1989), vol. 12, pp. 437–41.

1990

"The Evidence for Early European Angling, II: The Mysterious Manuscript of Astorga, 1624," *The American Fly Fisher* 16:3 (1990), pp. 8–16.

1993

"The Oldest Use of Silk in Fly Fishing," *The American Fly Fisher* 19:3 (Winter 1993), pp. 16–19.

1994
"Remains and Verbal Evidence of Carp (*Cyprinus carpio*) in Medieval Europe," in *Fish Exploitation in the Past: Proceedings of the 7th Meeting of the I.C.A.Z. Fish Remains Working Group*, ed. Wim Van Neer (Tervuren, 1994), pp. 139–50.
"Medieval Cistercian Fisheries, Natural and Artificial," in *L'espace cistercien*, ed. Léon Pressouyre Ministère de l'Enseignement supérieur et de la Recherche, Comité des travaux historiques et scientifiques, Mémoires de la section d'archéologie et d'histoire de l'art 5 (Paris, 1994), pp. 401–14.
"The Craft of Fishing Alpine Lakes, ca. A.D. 1500," in *Archaeo-Ichthyological Studies: Papers presented at the 6th Meeting of the I.C.A.Z. Fish Remains Working Group, Schleswig*, ed. D. Heinrich, Beiheft zur *Offa* 51 (Neumünster, 1994), pp. 308–13.

1995
"Fishing," in *Medieval France: An Encyclopedia*, ed. W. Kibler and G. Zinn (New York, 1995), p. 347.
"The Evidence for Early European Angling, III: Conrad Gessner's Artificial Flies, 1558," *The American Fly Fisher* 21:1 (Spring 1995), pp. 2–11, with an addendum, *The American Fly Fisher* 21:2 (Summer 1995), p. 24; German translation by Günter Fröhlich: "Conrad Gessners künstliche Fliegen, 1558," *Der Fliegenfischer* 166 (February/März 2006), pp. 40–3.
"Environmental Change and the Culture of Common Carp in Medieval Europe," *Guelph Ichthyology Review* 3 (1995): 57–85; reprinted in *The Completely Illustrated Guide to Koi for Your Pond*, ed. H. R. Axelrod et al. (Neptune City, N.J., 1996), pp. 272–301.
"*Carpes pour le duc* … The Operation of Fish Ponds at La Perriére-sur-Saône, Burgundy, 1338–52," *Archaeofauna* [Madrid] 4 (1995), 33–45.
"Fishers in Late Medieval Rural Society Around Tegernsee, Bavaria – A Preliminary Sketch," in *The Salt of Common Life: Individuality and Choice in the Medieval Town, Countryside and Church. Essays Presented to J. Ambrose Raftis*, ed. Edwin B. Dewindt (Kalamazoo, 1995), pp. 371–408.

1996
"Economic Development and Aquatic Ecosystems in Medieval Europe," *American Historical Review* 101 (1996), 631–69; reprinted in *Internal Colonization in Medieval Europe*, ed. James Muldoon and Felipe Fernández-Armesto (Burlington, VT, 2008), pp. 123–61.

1997
*Fishers' Craft and Lettered Art: Tracts on Fishing from the End of the Middle Ages* (Toronto, 1997) [xvi + 403 pp.]; excerpted in *The American Fly Fisher* 24:1 (Winter 1998), pp. 1–10.
"Pike (*Esox lucius*) in Late Medieval Culture: From Illiterate Empiricism to Literate Traditions," *Internet Archaeology* 3 (Autumn 1997) [electronic publication].

1999

"Fish und Men: Changing Relations in Medieval Central Europe," *Beiträge zur Mittelalterarchäologie in Österreich* 15 (1999), 187–96.

2000

"Medieval Fishing," in *Working with Water in Medieval Europe: Technology and Resource Use*, ed. Paolo Squatriti (Leiden, 2000), pp. 331–93.

"Fishing and Fish Culture, Western Christian," *Encyclopedia of Monasticism*, ed. William M. Johnston (Chicago, 2000), pp. 475–7.

2001

"Frontier Foods for Late Medieval Consumers: Culture, Economy, Ecology," *Environment and History* 7 (2001), 131–67.

"A Longer View: Is Industrial Metabolism Really the Problem?" in *Nature, Society and History: Long Term Dynamics of Social Metabolism*, ed. M. Fischer-Kowalski, E. Rosa, R. P. Sieferle, and B. Smetschka, as a special issue of *Innovation: The European Journal of Social Sciences* 14 (2001), 143–55.

2002

"Carp, Cods, Connections: New Fisheries in the Medieval European Economy and Environment," in *Animals in Human Histories: The Mirror of Nature and Culture*, ed. Mary J. Henninger-Voss (Rochester, N.Y., 2002), pp. 3–55.

2003

"*Artorius Rex Britanniae* from a Contemporary Witness" in *Proceedings of the Pseudo Society: First Series, 1986–1993*, ed. Richard Ring and Richard Kay (Kalamazoo, 2003), pp. 175–9.

2004

"Ecology," in *Dictionary of the Middle Ages: Supplement 1*, ed. William C. Jordan (New York, 2004), pp. 169–74.

Editor, "Human Economy and Natural Environment in the Middle Ages [collected abstracts from sessions at Kalamazoo Medieval Congress, 2004]," *AVISTA Forum Journal* 14 (Fall 2004), pp. 62–8.

2005

"A Brief History of Aquatic Resource Use in Medieval Europe," in *Ecological History of the Wadden Sea*, ed. Heike K. Lotze and Karsten Reise, as a special issue of *Helgoland Marine Research* 59 (2005), 22–30.

(with Heike K. Lotze and others), "Human Transformations of the Wadden Sea Ecosystem through Time: A Synthesis," in *Ecological History of the Wadden Sea*, ed. Heike K. Lotze and Karsten Reise, as a special issue of *Helgoland Marine Research* 59 (2005): 84–95.

(with Bernd Hermann and others) "Abundance of Species in Historic Times: Report of an Ad-Hoc Research Group," in *History and Sustainability: Proceedings of the Third International Conference of the European Society for*

*Environmental History, Florence, Italy, February 16–19, 2005*, ed. Mauro Agnoletti et al. (Florence, 2005), pp. 147–9.
"Medieval Christendom in God's Creation: Environmental Continuities, Coevolutions, and Changes," in *Northern Europe: An Environmental History*, ed. Tamara Whited et al. (Santa Barbara, Denver and Oxford, 2005), pp. 45–72.
"Recommended Readings: A Sixteenth-Century Exemplar and Source," *Environmental History* 16 (2005), 689–91.
"Fishing," in *Medieval Science, Technology and Medicine: An Encyclopedia*, ed. Thomas F. Glick, Steven J. Livesey, and Faith Wallis (New York and London, 2005), pp. 175–6.

2006
"Der Karpfen (*Cyprinus carpio* L.): Der lange Weg eines 'Fremdlings' in die Schweiz," in *Fisch und Fischer aus zwei Jahrtausenden: Eine fischereiwirtschaftliche Zeitreise durch die Nordwestschweiz*, ed. Heide-Marie Hüster-Plogmann (Augst, Switzerland, 2006), pp. 161–7.
Editor, "Environmental History at Kalamazoo, 2006 [collected abstracts from sessions at Kalamazoo Medieval Congress, 2006]," *AVISTA Forum Journal* 16 (Fall 2006), 55–9.

2007
"Footprint Metaphor and Metabolic Realities: Environmental Impacts of Medieval European Cities," in *Natures Past: The Environment and Human History*, ed. Paolo Squatriti (Ann Arbor, 2007), pp. 288–325.
(with Dirk van Damme, Nina Bogutskaya, and Carl Smith) "The Introduction of the European Bitterling (*Rhodeus amarus*) to West and Central Europe," *Fish and Fisheries* 8 (2007), 79–106.

2008
(with four co-participants) "AHR Conversation: Environmental Historians and Environmental Crisis," *American Historical Review* 113 (2008), 1431–65.
"Aquaculture in Champagne before the Black Death of 1348–1350," in *Archéologie du poisson – 30 ans d'archéo-ichtyologie au CNRS: Hommage aux travaux de Jean Desse et Nathalie Desse-Berset*, ed. P. Béarez, S. Grouard, and B. Clavel, XXVIIe recontres internationales d'archéologie et d'histoire d'Antibes (Antibes, 2008), pp. 67–82.
Editor, "Environmental History at Kalamazoo, 2008 [collected abstracts from sessions at Kalamazoo Medieval Congress, 2008]," *AVISTA Forum Journal* 18 (Fall 2008), 98–102.
"*Homo et Natura, Homo in Natura*: Ecological Perspectives on the European Middle Ages," in *Nature in the Middle Ages and the Renaissance*, ed. Lisa Kiser and Barbara Hanawalt (South Bend, Indiana, 2008), pp. 11–38.
"Medieval Europeans and their Aquatic Ecosystems," in *Beiträge zum Göttinger Umwelthistorischen Kolloquium 2007–2008*, Graduiertenkolleg

Interdisziplinäre Umweltgeschichte, ed. Bernd Herrmann (Göttingen, 2008), pp. 45–64.
"*Strekfusz*: A Fish Dish Links Jagiellonian Cracow to Distant Waters," in *Central and Eastern Europe in the Middle Ages: A Cultural History*, ed. Piotr Górecki and Nancy van Dusen (London, 2008), pp. 116–24 and 261–5.

* * *

In addition to many forthcoming articles, Richard Hoffmann is currently completing two books that promise to have an important impact on the study of premodern environmental history: a monograph entitled *The Catch: Medieval European Fisheries and the Antecedents of Today's Global Fisheries Crisis – An Essay in Environmental History* (University of Pennsylvania Press); and a textbook entitled *An Environmental History of Medieval Europe* (Cambridge University Press).

# INDEX